An Introduction to Free Radical Chemistry

'Too many free radicals, that's your problem.'

'Free Radicals, Sir?'

'Yes. They're toxins that destroy the body and brain—caused by eating too much red meat and white bread and too many dry Martinis.'

'Then I shall cut out the white bread, Sir.'

James Bond, *Never Say Never Again* (1983)

An Introduction to Free Radical Chemistry

Dr A.F. Parsons
Department of Chemistry
University of York
York

**Blackwell
Science**

© 2000
Blackwell Science Ltd
Editorial Offices:
Osney Mead, Oxford OX2 0EL
25 John Street, London WC1N 2BS
23 Ainslie Place, Edinburgh EH3 6AJ
350 Main Street, Malden
  MA 02148-5018, USA
54 University Street, Carlton
  Victoria 3053, Australia
10, rue Casimir Delavigne
  75006 Paris, France

Other Editorial Offices:
Blackwell Wissenschafts-Verlag GmbH
Kurfürstendamm 57
10707 Berlin, Germany

Blackwell Science KK
MG Kodenmacho Building
7–10 Kodenmacho Nihombashi
Chuo-ku, Tokyo 104, Japan

The right of the Author to be
identified as the Author of this Work
has been asserted in accordance
with the Copyright, Designs and
Patents Act 1988.

First published 2000

Set by Graphicraft Limited, Hong Kong

DISTRIBUTORS

Marston Book Services Ltd
PO Box 269
Abingdon, Oxon OX14 4YN
(*Orders*: Tel: 01235 465500
        Fax: 01235 465555)

USA
Blackwell Science, Inc.
Commerce Place
350 Main Street
Malden, MA 02148-5018
(*Orders*: Tel: 800 759 6102
          781 388 8250
        Fax: 781 388 8255)

Canada
Login Brothers Book Company
324 Saulteaux Crescent
Winnipeg, Manitoba R3J 3T2
(*Orders*: Tel: 204 837 2987)

Australia
Blackwell Science Pty Ltd
54 University Street
Carlton, Victoria 3053
(*Orders*: Tel: 3 9347 0300
        Fax: 3 9347 5001)

A catalogue record for this title
is available from the British Library

ISBN 0-632-05292-9

Library of Congress
Cataloging-in-publication Data

Parsons, A.F.
    An introduction to free radical chemistry/
    A.F. Parsons.
        p.   cm.
    Includes bibliographical references and index.
    ISBN 0-632-05292-9
    1. Free radicals (Chemistry)   I. Title.
    QD471 .P36 2000
    541.2'24—dc21

For further information on
Blackwell Science, visit our website:
www.blackwell-science. com

Contents

CONTENTS

Chemical Abbreviations

Ac	acetyl (ethanoyl)	CH_3CO-
Bn	benzyl	$C_6H_5CH_2-$
BOC	tertiary-butoxycarbonyl	$(CH_3)_3CCO_2-$
Bu	butyl	$CH_3CH_2CH_2CH_2-$
tBu	tertiary-butyl	$(CH_3)_3C-$
Bz	benzoyl	C_6H_5CO-
Et	ethyl	CH_3CH_2-
Me	methyl	CH_3-
Ms	methanesulfonyl	CH_3SO_2-
Ph	phenyl	C_6H_5-
Ts	toluenesulfonyl	$4\text{-}CH_3C_6H_4SO_2-$

Ar	aryl
E	electrophile
hv	irradiation
Nu	nucleophile
R	alkyl, aryl etc.
X, Y	leaving groups (mostly halogens)

Preface

It is only within the last 20–30 years that the considerable importance of free radicals (or radicals) in both chemistry and biology has really come to light. Research has shown, for example, that radicals are present in the atmosphere, in our bodies and during some very important chemical reactions. On the hundredth anniversary of the work of Gomberg, which established the existence of a carbon-centred radical, it is now remarkable to see how radical chemistry impinges on many aspects of our daily lives. Examples range from the destruction of the protective ozone layer in the stratosphere, to the large-scale preparation of plastic products used in the household.

This book aims to give an insight into the world of radical chemistry and, in particular, organic radical chemistry. It was inspired by the considerable number of students who, after my lecture course, would complain about the lack of a suitable textbook to consolidate and expand their understanding of free radical chemistry. As a consequence, I decided to put pen to paper and write this book, which is aimed at providing a clear, structured (and hopefully interesting!) introduction to radicals and especially to their use in organic synthesis. This is by no means a comprehensive account, but it is intended to provide a gentle, stepwise approach to understanding the key principles involved in radical reactions. The intended audience is advanced second, third and fourth year chemistry and biochemistry (and related) undergraduates. It is also appropriate for postgraduates, academics and industrialists who need an introduction or need to brush up on their knowledge of radical chemistry.

The book begins by defining what a radical is and then introduces some of the most important applications of radical reactions in industry and in nature. Chapter 2 discusses radical structure and stability, and this is followed by an outline of the various methods used to prepare radicals in Chapter 3. The main pathways by which radicals react are then discussed in Chapter 4, and this is followed, in Chapter 5, by an overview of the factors which need to be considered when designing new radical reactions. This includes a comparison of radical reactions with (traditional) ionic reactions. At this stage, the fundamental principles of radical formation and reactivity are discussed and Chapters 6–9 build on these principles by illustrating important synthetic applications. Hence, functional group transformations, intramolecular cyclizations, intermolecular reactions and radical translocation reactions are all covered individually. Case studies are included to illustrate these reactions in 'landmark' syntheses. Finally, Chapters 10 and 11 discuss the formation and reactivity of radical anions and cations, which undergo reactions characteristic of both radicals and ions. A summary is included at the end of each chapter to emphasize key points and questions are provided to test the reader's understanding. Outline answers to the questions are included, together with references to the original literature. References are also included within the later chapters, and a further reading section is provided at the end of each chapter and at the end of the book so that the reader can access more detailed information.

There are numerous people I must thank for their help with this project. These include many research students and colleagues at York (including Peter Hanson, John Lindsay Smith, Barry Thomas, John Vernon) and, in particular, Bruce Gilbert and Adrian Whitwood. Their enthusiasm and constructive comments were invaluable. I would also like to thank my family for their support; it was only with the tremendous support and patience of Sue and Thomas (and the menagerie) that the book was written.

Finally, if this book encourages a student to choose (and do well on!) a radical question in an examination, or a research student to study a project on radical chemistry, or an industrialist to think about using radical reactions in the synthesis of target molecules, then I have succeeded.

Dr Andrew Parsons
2000

Radicals and their Importance

1.1 What are radicals?

Radicals have an impact on all of our lives. We make radicals in our bodies, radicals are produced when we light a fire or drive a car, and we use many products (chiefly plastics) that are produced on a large scale using radical reactions. Radicals affect our health and, for example, a leading explanation for the ageing process places the blame on harmful radicals. Destructive radicals also affect our environment and radicals generated from chlorofluorocarbons (CFCs) are responsible for the destruction of the Earth's protective ozone layer. So, what are radicals?

We can define radicals as atoms or compounds which contain an unpaired electron. They all contain an odd number of electrons. Simple examples include hydrogen and bromine atoms ($H^\bullet$ and $Br^\bullet$), which have one and seven electrons, respectively, in their outer valence shell (Fig. 1.1). The single unpaired electron for each atom is represented as a dot. Almost all radicals can be described as 'free radicals' as they exist independently, free of any support from other species. They are generally very unstable and are regarded as reactive intermediates, together with carbocations (R_3C^+), carbanions (R_3C^-) and carbenes ($R_2C{:}$). This high reactivity is due to the unpaired electron which would like to 'pair' with a second electron to produce a filled outer shell. Thus, the outer electrons in the hydrogen and bromine atoms could pair and produce hydrogen bromide (Fig. 1.1), or two hydrogen or two bromine atoms could combine (dimerize) to give H_2 and Br_2, respectively. The driving force in each case is the formation of a two-electron covalent bond. If we wanted to prepare hydrogen and bromine radicals, we could consider the reverse process and attempt to break the covalent bond of H—Br by applying energy (e.g. heat or light).

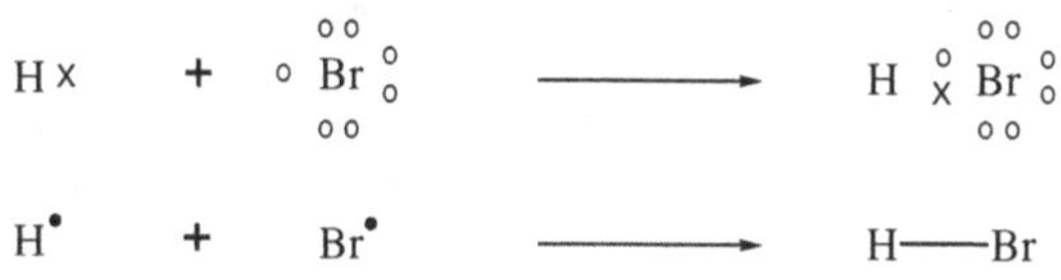

Fig. 1.1

This type of bond cleavage, to give each atom one electron, is known as homolysis or homolytic bond cleavage.

There are various pathways by which radicals can react to form stable molecules. They can combine with themselves or other radicals, but they can also be oxidized to a cation (by loss of an electron) or reduced to an anion (by addition of an electron). These ions could then react with nucleophiles or electrophiles, respectively, to produce neutral and stable products. This is illustrated for a carbon-centred radical in Fig. 1.2. These radicals contain seven valence electrons, which is one electron more than carbocations and one electron less than carbanions. We can see that both the radical and cation are electron deficient as they require the addition of one and two electrons, respectively, to produce a filled (eight-electron) outer shell.

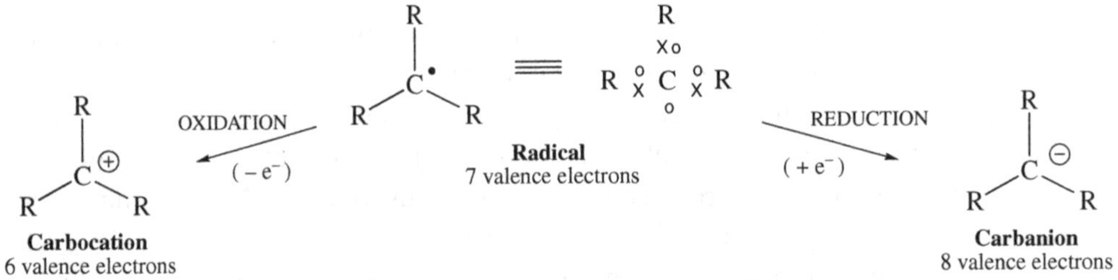

Fig. 1.2

Not all radicals are highly reactive. Notable, naturally occurring exceptions include oxygen and nitrogen monoxide ($^\bullet$NO). Molecular oxygen (O_2) can be thought of as a di- or biradical as it contains two unpaired electrons and, for simplicity, this will be represented as $^\bullet$O—O$^\bullet$.[1] Whereas radicals have an odd number of electrons, biradicals have an even number and O_2 has 12 electrons.

This introductory chapter will now outline a brief history of radical chemistry and go on to discuss a variety of biologically and industrially important radical reactions.

1.2 The discovery and development of radical chemistry

Lavoisier first used the term 'radical' in 1789 when he described acids as being composed of oxygen and an entity called a radical. Even after work

---

[1] It should be noted that molecular orbital theory predicts that O_2 has a triplet ground state with a double bond.

by Davy, which showed that compounds can be acids without containing oxygen, the term persisted. Organic chemists introduced a number of names to help categorize organic groups and, for example, the CH_3 and CH_3CH_2 groups were known as methyl and ethyl radicals. However, the problems of determining molecular formulae resulted in confusion. For example, in 1849, Kolbe described the product derived from the electrolysis of potassium acetate (ethanoate) as 'methyl radical' with a formula of C_2H_3. We now know that the product is ethane (C_2H_6) which is actually produced from the reaction (dimerization) of two methyl radicals ($^{\bullet}CH_3$)!

In 1847, Faraday first demonstrated that oxygen is drawn into a magnetic field, and hence is strongly paramagnetic, while nitrogen monoxide is weakly paramagnetic. We now associate paramagnetism with molecules (or ions) which contain unpaired electrons, as the spinning electron behaves like a tiny magnet and the molecule is drawn into a magnetic field.

Victor Meyer then showed that an iodine molecule (I_2) could dissociate into iodine atoms or radicals ($I^{\bullet}$). However, the key breakthrough came in 1900 when Gomberg investigated the reaction of triphenylmethyl bromide (1) with silver (Fig. 1.3). In the absence of oxygen, the reaction yielded a highly reactive white solid which, when dissolved, gave a yellow solution. Gomberg proposed that the product was hexaphenylethane (3) which, when in solution, existed in equilibrium with the coloured triphenylmethyl radical (2). The presence of a radical helped to explain why other radicals, including oxygen, reacted rapidly with the product. This was a major discovery and chemists started to believe that radicals could actually exist. By 1911, the experimental evidence for (2) was so overwhelming that the case for free radicals was firmly established. So, why was Gomberg able to observe this particular radical? Part of the reason for the stability of (2) lies with the presence of three bulky benzene rings. These effectively shield the central carbon atom bearing the radical and slow down any reactions. On the basis of this argument, we would expect that (3), which is highly crowded, would be difficult to form. This is indeed the case, and we now know that the combination of two radicals of type (2) gives the less strained product (4). This was the product actually isolated by Gomberg and we will look at how this is formed in Section 2.4.2.

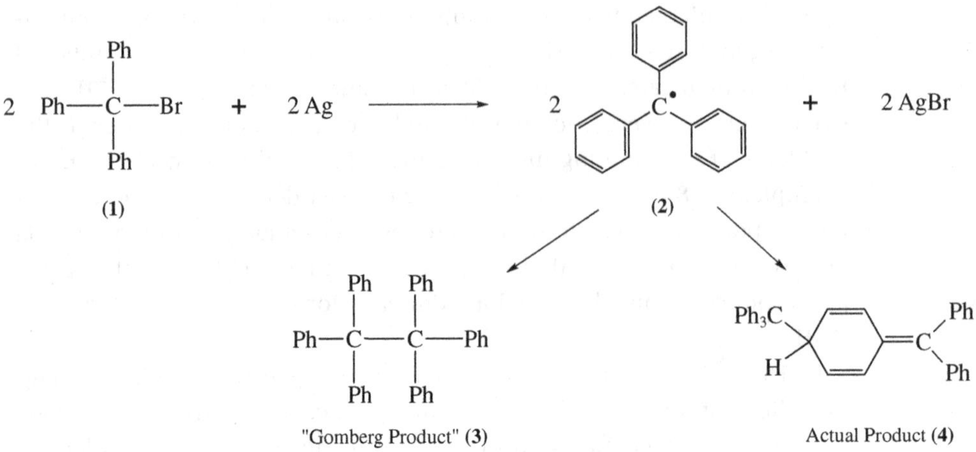

Fig. 1.3

The formation of less stable, more reactive, radicals then followed. In 1929, Paneth showed that tetramethyllead produces a metallic lead mirror when heated at high temperatures (around 200°C) in a glass tube containing a stream of unreactive carrier gas. Experiments showed the momentary existence of methyl radicals (in the gaseous phase) which could either combine with themselves, to give ethane, or remove the lead mirror to regenerate starting material (Fig. 1.4). Similar results were obtained using tetraethyllead (PbEt$_4$), and homolytic cleavage of the Pb—C bond gives ethyl radicals and lead. These compounds have since found a use as antiknock agents in petrol engines, as the radicals which are generated make the fuel burn more smoothly (see Section 1.4).

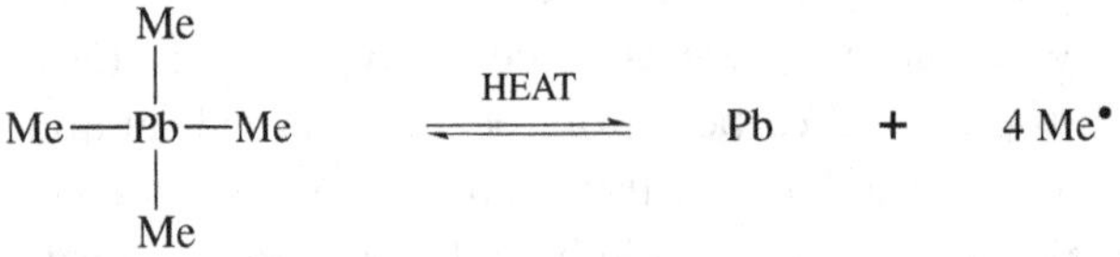

Fig. 1.4

It was not until 1937 that radicals were postulated to be intermediates in a variety of chemical reactions. Mechanisms proposed by Hey and Waters to explain the formation of biphenyls (from aromatics and benzoyl peroxide) and Kharasch to explain the orientation of addition of HBr to unsymmetrical alkenes (in the presence of peroxides) both included radical intermediates. Both processes involved the use of peroxides (RO—OR).

This is no coincidence as the first step in these reactions involves cleavage of the weak oxygen–oxygen bond to generate oxygen-centred radical (RO$^{\bullet}$) intermediates. Indeed, the term 'peroxide effect' is often used to account for the anti-Markownikoff (Markovnikov) addition of HBr to alkenes. This work also demonstrated the concept of radical chain reactions in which an initial radical A$^{\bullet}$ reacts to form a second radical B$^{\bullet}$ which, in turn, reacts to give radicals C$^{\bullet}$, D$^{\bullet}$, etc. The process is terminated by the combination of two radicals, for example A$^{\bullet}$ and B$^{\bullet}$, to give non-radical products, which in this case would be A—B.

This was followed by further investigations into the reaction of radicals with alkenes to give polymers. These products were targeted as potential substitutes for natural rubber, which was unavailable to the Allied Forces during the Second World War. It was during this time that Mayo, Walling and Lewis investigated radical polymerizations using two different alkene monomers (so-called copolymerization). Monomers A and B could react to give an alternating polymer of the type ABABABAB, etc., and these results were explained by considering the intermediate radicals. They showed that not all carbon-centred radicals are the same; they have different 'characters' and as a result can react differently. This is an important concept as it allows us to explain and also predict selective radical reactions.

It was not until the 1950s to 1970s that physical organic chemists began to quantify radical reactions and determine absolute rates of reaction in solution. This was revolutionized by the development of a new technique, known as electron spin resonance (ESR) spectroscopy, which offered a very sensitive method for detecting and identifying even short-lived radicals. The formation and disappearance of a variety of radicals could now be monitored, and important information on the structure of the radicals could be obtained.

It was also around this time that researchers began to investigate and propose mechanisms for the deterioration of fats, oils and other foodstuffs in the presence of oxygen. Radical intermediates were shown to be involved, and the term autoxidation was introduced to describe this process. This prompted the design of molecules, called antioxidants or inhibitors, which could slow down or even stop these undesirable reactions.

With a knowledge of reaction rates, chemists could now start to explore the use of radicals in the preparation of not only polymers but also small molecules. Since 1970, a number of important radical reactions have been developed and numerous target molecules have been prepared using these methods. Today, when planning the synthesis of even a complex molecule, chemists can make use of efficient and selective radical reactions that can have a number of advantages over more traditional ionic methods (involving anions and cations).

It should be emphasized that radical reactions do not only occur in research laboratories. The process by which benzaldehyde is oxidized in air to benzoic acid on a laboratory bench can be thought of as the same type of radical reaction as that which leads to the deterioration of foods, the ageing of unprocessed natural rubber and the drying of paints and varnishes. The following sections aim to show the considerable scope and relevance of radical chemistry in our everyday lives.

1.3 Natural radical reactions

Excessive exposure to environmental pollution (e.g. exhaust fumes), ultraviolet light or cigarette smoke, and illness, can cause the body to produce harmful radicals. It has been estimated, for example, that 10^{14} radicals (which include $^{\bullet}NO$ and $NO_2^{\bullet}$) are present in one puff of cigarette smoke. Left unchecked, destructive radicals can lead to a number of diseases in humans, including arthritis, cancer and Parkinsonism. Radicals may also start the damage that causes fatty deposits in the arteries, leading eventually to heart disease or a stroke, and experimental work points to the role of radicals in bovine spongiform encephalopathy (BSE).

A role for radicals in ageing was first suspected in the 1950s, and recent research attributes this to oxygen radicals derived from mitochondria. Mitochondria are the energy factories of cells where oxygen and nutrients are used to prepare adenosine triphosphate (ATP)—the molecule that powers most other activities in cells. Unfortunately, one by-product of this process is the superoxide radical anion ($O_2^{-\bullet}$), a charged radical, which, in turn, can be converted to a hydroxyl radical ($HO^{\bullet}$), the most reactive oxygen-centred radical known. The hydroxyl radical is particularly indiscriminate in its choice of reactant and can damage proteins, fats or deoxyribonucleic acid (DNA) within the cell. These reactions can interfere with the proper functioning of the cell, leading to its death and, ultimately, that of the organism. Radicals of this type are also produced when water is irradiated with high-energy radiation (e.g. ultraviolet light, X-rays or γ-radiation), and this is important to us as water accounts for 50–60% of our body weight. Exposure to sunlight, X-radiation or nuclear radiation can lead to the formation of these destructive radicals in our bodies which, in turn, can promote cancers, sterility and even death. Hydroxyl radical generation is actually used beneficially to treat cancers in radiotherapy; the rapidly multiplying cancer cells are killed by exposure to a radioactive material (e.g. ^{60}Co) or X-rays.

It is not all doom and gloom, though, as our body is not defenceless—living cells can produce enzymes which detoxify free radicals, including superoxide dismutases (SODs). These are metal-containing enzymes (e.g.

manganese, copper or zinc) which convert $O_2^{-\bullet}$ to oxygen and hydrogen peroxide (H_2O_2). The reactive H_2O_2 is rapidly reduced to water by catalase or peroxidase enzymes. β-Carotene (or provitamin A), vitamin C (ascorbic acid) and vitamin E (α-tocopherol) can also delay or inhibit oxidative damage, and these types of molecules are thus known as antioxidants. Vitamin E is fat soluble and protects against radical damage within the cell membrane; laboratory experiments suggest that this process also involves the water-soluble vitamin C. These molecules scavenge radicals, and vitamin E can react with an alkoxyl radical by donating a hydrogen atom (from the OH group) to give an alcohol and a vitamin E radical (Fig. 1.5), which is much less harmful. Vitamin E is believed to be re-generated by reaction with vitamin C at the cell membrane (water–lipid) surface.

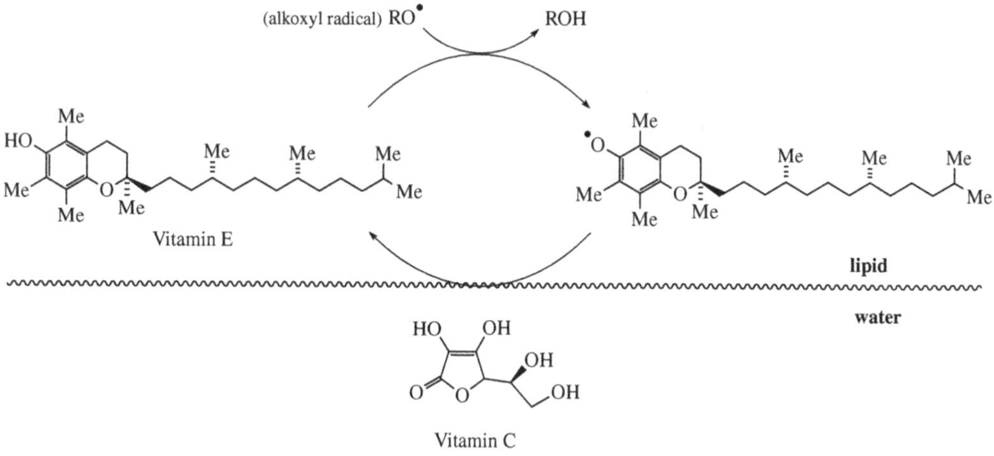

Fig. 1.5

Our antioxidant defences therefore rely on both vitamins and minerals, from which metal-containing enzymes are prepared, in our diet. A variety of alternative antioxidants are present in foods and include flavonoids, such as quercetin (Fig. 1.6). Flavonoids are a group of around 1000 compounds found in certain fruit and vegetables, including citrus fruits, apples and grapes. Tea leaves also contain flavonoids (and related polyphenols), while garlic produces a powerful antioxidant called allicin (Fig. 1.6). Just how important these dietary antioxidants are to our health is a matter of debate. Recent research suggests, however, that antioxidants may offer protection against cancers and heart disease. So, perhaps the antioxidants in apples really do keep the doctor away!

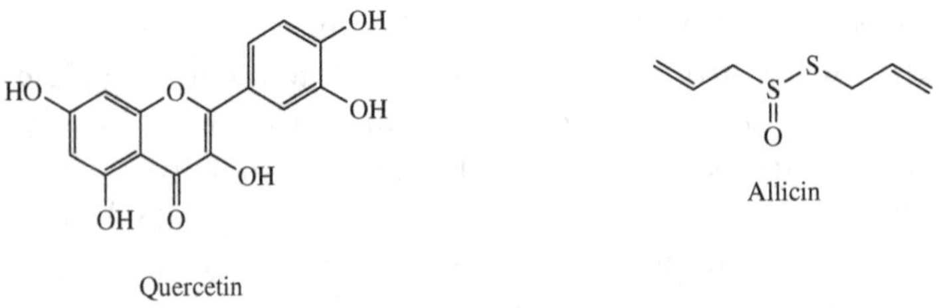

Quercetin

Allicin

Fig. 1.6

Not all radical reactions involving molecular oxygen are destructive in humans. A very important series of beneficial reactions arises from the reaction of oxygen with an unsaturated acid called arachidonic acid (Fig. 1.7). These radical reactions, often called the 'arachidonic acid cascade', produce a variety of biologically active compounds (including prostaglandins, thromboxanes and leukotrienes) which are vital for the body to function properly. Of particular medicinal importance are the prostaglandins, including prostaglandin $F_{2\alpha}$, which exhibit a range of important activities, including contraction of smooth muscle, regulation of cell function and fertility, lowering of arterial blood pressure and control of platelet aggregation.

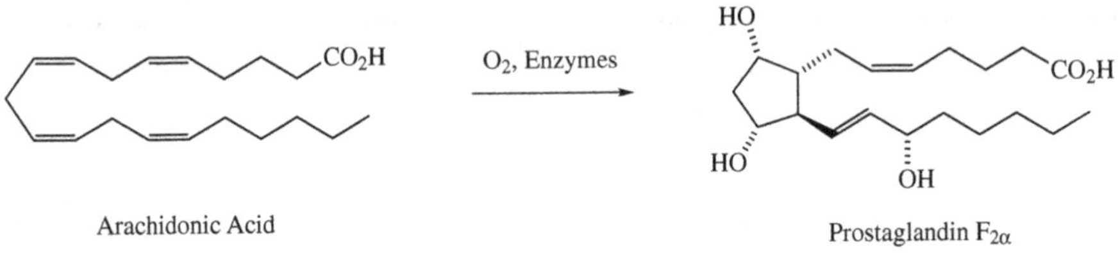

Arachidonic Acid   Prostaglandin $F_{2\alpha}$

Fig. 1.7

One final example of natural radical reactions involves the enediyne family of compounds isolated from various bacteria. These are some of the most potent antitumour agents ever discovered and, for example, dynemicin A (Fig. 1.8) is known to prolong significantly the lifespan of mice inoculated with leukaemia cells. The remarkable biological activity of these compounds results from their reaction with DNA in the rapidly multiplying cancer cells. Enediyne antitumour agents are able to generate a very reactive biradical intermediate (typically a 1,4-benzyne), known as the 'warhead', that can selectively abstract two hydrogen atoms from the DNA backbone. This leads to irreversible cleavage of the DNA strand which, in turn, leads to cell death.

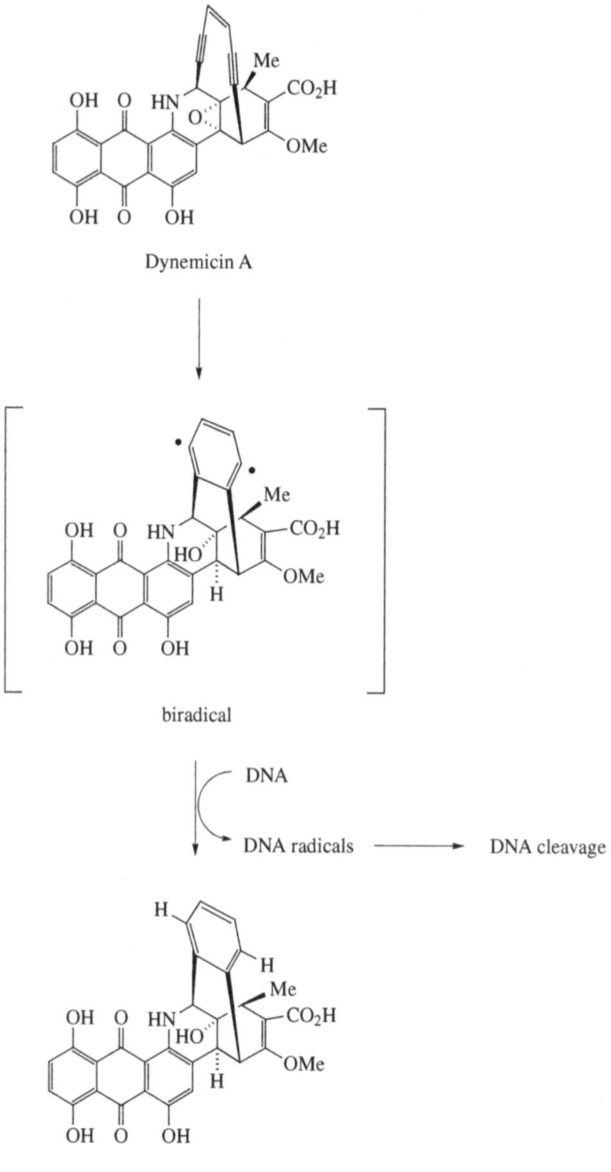

Dynemicin A

biradical

DNA ⟶ DNA radicals ⟶ DNA cleavage

Fig. 1.8

1.4 Commercial radical reactions

The most important industrial radical reactions are used for the manufacture of polymers. Around 10^8 tonnes (or 75%) of all polymers are prepared using radical processes. These are chain reactions in which an initial radical adds to the double bond of an alkene monomer and the resulting radical adds to another alkene monomer, and so on. This 'addition' polymerization is used to make a number of important polymers, including poly(vinyl chloride) (PVC), polystyrene, polyethylene (polythene) and poly(methyl methacrylate) (Fig. 1.9). Copolymers (see earlier) can also be easily prepared starting from a mixture of two or more monomers. These polymers have found widespread use as they possess a range of chemical and mechanical (i.e. strength, toughness) properties.

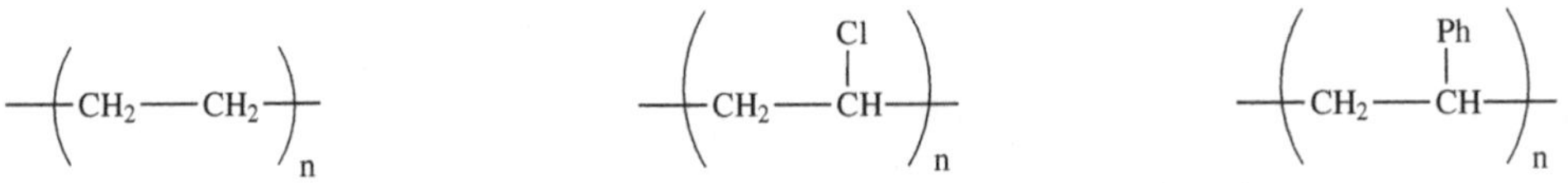

$-(CH_2-CH_2)_n-$	$-(CH_2-CH_{Cl})_n-$	$-(CH_2-CH_{Ph})_n-$
Polyethylene (bottles, tubing, sheets)	Poly(vinyl chloride) (raincoats, shower curtains, garden hose, rigid clear bottles, swimming pool liners)	Polystyrene (moulded objects, electrical insulation)

Fig. 1.9

A related polymerization reaction is responsible for the drying of paints, varnishes and oils. Oil-based varnishes contain mixtures of long-chain unsaturated fatty acids such as linoleic acid (Fig. 1.10). When exposed to air, the acids react with oxygen to form various oxygen and carbon radicals. These can react with the double bonds of nearby molecules to 'cross-link' (or join) the acids, and this eventually leads to the formation of a solid polymer. Although these paints are cheap, they are slow drying and, unfortunately, the surface continues to react with oxygen so becoming yellow and brittle—a process known as weathering. More modern emulsion (or latex) paints contain a suspension of the polymer in water which quickly evaporates to leave a coating.

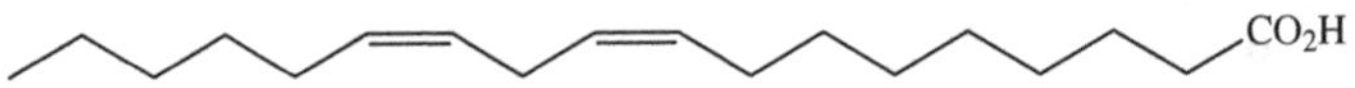

Linoleic Acid

Fig. 1.10

Combustion is a very important radical process that we encounter every day. The car engine is a good radical chemist: it exploits the reaction of petrol with molecular oxygen at high temperature. Although the mechanism of combustion of hydrocarbons is complex, many of the steps involve the production and reaction of carbon- and oxygen-centred radicals. If the production of these radicals is not checked, an explosion or 'knock' can occur. To prevent this problem, an anti-knock agent, such as $PbEt_4$ or, more recently, methyl *tert*-butyl ether ($MeOCMe_3$), is added to petrol. $PbEt_4$ is a source of ethyl radicals and these can combine with the hydrocarbon radicals stopping any further radical reactions and preventing an explosion.

Similar hydrocarbon radicals are generated in industrial plants devoted to 'cracking'. The high temperatures ($\geq 600°C$) ensure that radicals are produced from homolytic cleavage of the strong carbon–carbon and carbon–hydrogen bonds. This process allows valuable alkenes (such as ethene) and short-chain hydrocarbons to be prepared by degrading long-chain hydrocarbons extracted from crude oil.

Molecular oxygen can also oxidize a variety of organic compounds, including hydrocarbons, aldehydes, amines, ethers and ketones. These autoxidation reactions can be used to make a variety of small molecules, and a number of industrial processes rely on the controlled oxidation of organics using molecular oxygen (often with a metal catalyst). Examples include the formation of phenol and propanone (acetone) from cumene (isopropylbenzene), and cyclohexanone from cyclohexane (Fig. 1.11). Phenol is a popular starting material for a number of other products including aspirin and phenolic resins, whilst cyclohexanone can be oxidized (using nitric acid) to adipic acid, a precursor of nylon 6.

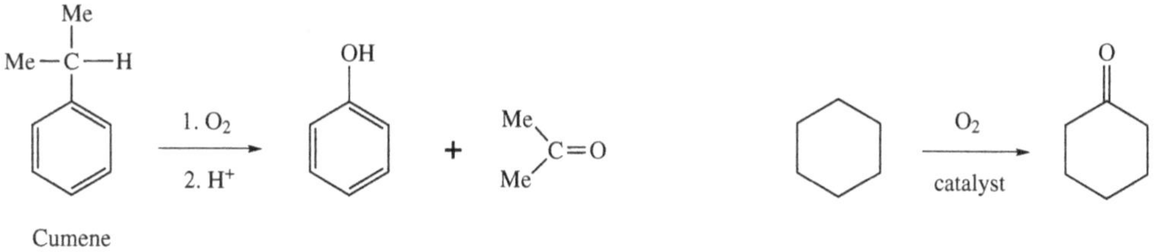

Fig. 1.11

Unfortunately, not all radical reactions of oxygen with organics are desirable. A notable example is reaction with unsaturated fatty acids; oxygen can react with the double bonds to produce ultimately a complex mixture of volatile, rancid-smelling, short-chain carboxylic acid products. (The same acids are present in human perspiration, and female dogs produce these compounds to attract a partner when on heat!) This autoxidation process can lead to rancid butters and particularly margarines which are rich in polyunsaturates. However, the radical chain reactions involved in the oxidation process can be prevented by sealing foodstuffs under nitrogen (e.g. potato crisps) or adding antioxidants, such as vitamin E, to foods. Other common antioxidants added to foods include butylated hydro-xyanisole (BHA) or E320 and butylated hydroxytoluene (BHT) (Fig. 1.12). These compounds have at least one bulky CMe_3 substituent, together with an OH group on the benzene ring, and both of these functional groups are necessary for their action as antioxidants. BHT is also added to natural rubber, which is mainly *cis*-polyisoprene (Fig. 1.12), to stop 'ageing' caused by radical oxidation reactions.

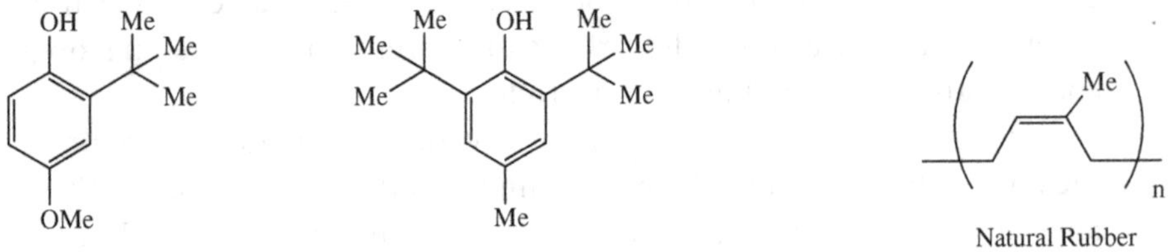

BHA   BHT   Natural Rubber

Fig. 1.12

This section can be concluded by mentioning a more recent develop-ment. This involves the disposal of chemical pollutants—an active area of research. One approach being investigated exploits the high reactivity of the hydroxyl radical generated from the ultraviolet irradiation of H_2O_2. These reactive radicals have been shown to break down a number of organic pollutants to harmless end-products. Further research will show if this method will be able to compete with more established methods including combustion.

1.5 Summary

Radicals are reactive intermediates formed by the homolysis of chemical bonds—the two-electron bond is broken to give one electron to each atom. They have an odd number of electrons and can combine, in order to 'pair' the electrons, to give stable non-radical products. Reaction with non-radical molecules, including alkenes, generates new radicals which can react further to form, for example, polymers in a chain reaction. Most, but not all, radicals are very reactive. Molecular oxygen, for example, is a stable biradical containing two unpaired electrons which slowly oxidizes organic compounds at moderate temperatures in autoxidation reactions.

Further reading

Donnelly, T.H. (1996) The origins of the use of antioxidants in foods. *Journal of Chemical Education*, **73**, 158–161.

Gutteridge, J.M.C. & Halliwell, B. (1994) *Antioxidants in Nutrition, Health and Disease*. Oxford University Press, New York.

Halliwell, B. & Gutteridge, J.M.C. (1999) *Free Radicals in Biology and Medicine*. Oxford University Press, New York.

Ochial, E.-I. (1993) Free radicals and metals in biology. *Journal of Chemical Education*, **70**, 128–133.

Seymour, R.B. (1988) Polymers are everywhere. *Journal of Chemical Education*, **65**, 327–334.

Walling, C. (1987) Fifty years of free radical chemistry. *Chemistry in Britain*, **23**, 767–770.

CHAPTER 2

The Basics

2.1 Introduction

We saw in Chapter 1 that radicals are formed by homolysis (or homolytic cleavage) of bonds. The two bonding electrons are divided equally, and this can be represented by one-electron, single-headed (or fish-hook) curly arrows (Fig. 2.1).

X—Y $\longrightarrow$ X$^\bullet$ + Y$^\bullet$ X$^\bullet$ $\longrightarrow$ X—Y

Fig. 2.1

The two single-headed arrows are drawn pointing in opposite directions. When a bond is formed by combination (or coupling) of two radicals, the arrows are drawn pointing towards each other.

This can be compared to heterolytic cleavage in which both bonding electrons are donated to *one* atom leading to the formation of ions (Fig. 2.2). One double-headed arrow is used to represent the movement of two electrons leading to the formation of an anion and a cation.

X—Y $\longrightarrow$ X$^\oplus$ + Y$^\ominus$ H—Cl $\longrightarrow$ H$^\oplus$ + Cl$^\ominus$

Fig. 2.2

Whereas single-headed arrows can point towards one another (resulting in the formation of a bond), double-headed arrows can never do this. The relative stability of the ions formed determines the direction of the double-headed arrow. In the example above, the Y group is able to stabilize the negative charge more effectively than the X group; this can be predicted from their electronegativities. For example, chlorine is more

14

electronegative than hydrogen,[1] and in HCl the arrow moves in the direction of the chlorine forming a chloride anion (together with a proton).

Heterolytic cleavage of bonds occurs at room temperature in polar solvents (which have high dielectric constants) as the ions are solvated and stabilized. In aqueous solution, HCl dissociates to H^+ and Cl^- ions which are both surrounded by a shell of water molecules; a lone pair of electrons on the oxygen (H_2O:) interacts with the proton, while hydrogen bonding stabilizes the chloride anion. Therefore, although these (oppositely) charged ions have a strong electrostatic interaction, the solvent shell slows down their recombination and it is possible to have very high concentrations of ions in polar solvents.

In the absence of polar solvents, however, the lowest energy pathway for breaking a single bond is not heterolysis, but homolysis. For example, when gaseous HCl is heated to above 200°C, the bond is broken homolytically to give hydrogen and chlorine atoms. The carbon–carbon bond of hydrocarbons can also be broken (homolytically) on heating at very high temperatures, and this is used in the oil refining industry to produce small-chain hydrocarbons (such as petrol) from crude oil (see Section 1.4). Radicals can also be generated in solution but, unlike ions, they are not charged and therefore they have very little interaction with the solvent. This is because no strong solvent shell is formed and most radicals will usually recombine very quickly (on almost every collision) to form non-radical products. This means that it is normally very difficult to generate high concentrations of radicals, and this restricts the methods that can be used for their detection.

2.2 Radical detection and observation

Indirect evidence for the formation of radicals in a reaction can often be obtained using chemical methods. In Chapter 1, we saw that radicals can add to alkenes leading to the formation of polymers. Therefore, if we were investigating the reaction of A + B → C, where radicals were suspected to be formed from A and/or B, we could add an alkene such as styrene ($PhCH{=}CH_2$) to the mixture of reactants. Any radicals generated in the reaction could then react to form polystyrene rather than product C. Alternatively, we could add radical inhibitors (or terminators) which, when used in small quantities, are known to prevent (or inhibit) radical reactions. These molecules react with radicals to give stable (unreactive)

---

[1] The Pauling scale of electronegativity gives values of 3.0 and 2.1 for Cl and H, respectively. The higher the value, the more electronegative the atom (e.g. F = 4.0 and Cs = 0.7).

products, thereby inhibiting the conversion of A and B to C. However, these methods, which are only appropriate for some reactions/radical intermediates, provide only limited information as they do not observe the radicals directly.

Perhaps the most obvious way to study radicals is to measure their magnetic moment. All radicals are paramagnetic, having one or more unpaired electrons, and this produces a net magnetic moment. They are drawn into a strong magnetic field and the magnetic susceptibility can be measured using a Gouy balance. This records the change in weight when a sample is placed in and out of a magnetic field. Unfortunately, this method is extremely insensitive and, in addition, most radicals are very reactive and have very short lifetimes. Infrared spectroscopy and mass spectrometry can be used to provide structural information for a very limited range of radicals. Ultraviolet (UV) spectroscopy can be used—the radical $(R^\bullet)$ often absorbs at higher wavelengths (λ_{max}), with an increased molar extinction coefficient or molar absorptivity (ε), than the non-radical precursor (RX) or products (e.g. RR)—but only the most stable radicals can be observed without special techniques. For example, pulse photolysis has been used, in which a laser generates a high concentration of radicals (in a very short time interval) and the UV spectrum is recorded. However, by far the most important methods for the detection of radicals are electron spin resonance (ESR) spectroscopy and, to a lesser extent, chemically induced dynamic nuclear polarization (CIDNP).

2.2.1 Electron spin resonance spectroscopy

ESR, also known as electron paramagnetic resonance (EPR), is an important spectroscopic technique for studying molecules (and ions) containing one or more unpaired electrons. It is a sensitive method and radicals can usually be detected at very low concentrations (down to 10^{-8} mol dm^{-3}). Analysis of the spectra can provide information on both the identity and structure of the radical. As a result, this powerful method has been used to help solve the reaction mechanisms involved in a multitude of chemical and biological processes.

ESR spectroscopy is very similar to proton nuclear magnetic resonance (^{1}H NMR) spectroscopy and, when an electron is placed in a magnetic field, it can align its moment with or against the field. The two spin states of the electron, $m_s = +1/2$ and $m_s = -1/2$, like the nuclear spin states $I =$

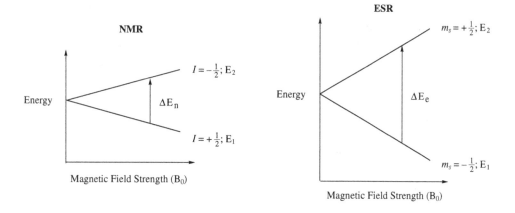

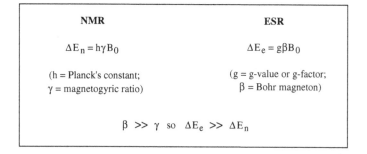

Fig. 2.3

$+1/2$ and $I = -1/2$ for hydrogen, are split apart by the applied magnetic field (Fig. 2.3). When irradiated with microwave radiation of the appropriate frequency (typically 10 GHz for a magnetic field of 3500 G), electrons in the lower energy state (E_1) can be promoted to the upper level (E_2). The frequency of the radiation which is absorbed corresponds to the energy difference (ΔE_e for an electron and ΔE_n for a ^1H nucleus). This gives rise to a net absorption signal, of low intensity, because absorptions can only occur when more electrons are in the lower level. As the energy difference, ΔE_e, is small (typically 3.8 J mol^{-1}), the excess population of the lower level is only around 1 in 700 electrons at room temperature. Like ΔE_n in NMR, the size of ΔE_e is proportional to the magnetic field strength (B_0) and, at equivalent field strengths, the splitting of the $m_s = \pm 1/2$ energy levels is larger in ESR than the $I = \pm 1/2$ energy levels in NMR. This is because the Bohr magneton, or magnetic moment for an unpaired electron, is much larger than the magnetic moment of a nucleus (measured by the magnetogyric ratio). ESR spectrometers can therefore be operated by less powerful magnets than can NMR instruments.

In order to improve the intensity (or the signal to noise ratio) of the absorption signal, the spectrum is generally recorded as the first derivative (the gradient of the absorption curve), as opposed to the direct absorption curve, which is the conventional way of presenting NMR spectra (Fig. 2.4). The different shape of the first derivative curve also allows better separation of two overlapping peaks (which are often seen in ESR spectra).

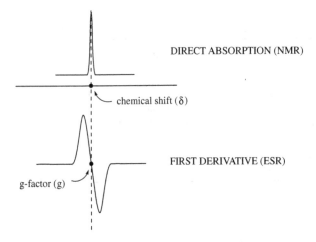

Fig. 2.4

In ^{1}H NMR spectroscopy, the different environments of hydrogen atoms in a molecule lead to slightly different energies of absorption. These are recorded and the different absorption signals give rise to 'chemical shifts' (in δ ppm); these are measured relative to an internal standard, usually tetramethylsilane (Me_4Si, TMS). For ESR spectroscopy, we replace chemical shifts with the term 'g-factor' (or g-value); different values of the g-factor correspond to different chemical shifts. However, the g-factor usually varies little from one radical to another. For carbon-centred radicals, with one free electron, the g-factor is always close to 2.0 (or, to be more precise, 2.0023). Therefore, whereas for ^{1}H NMR spectroscopy we have a chemical shift range of around 0–10 ppm, for carbon radicals there is a range of $g = 2.00$–2.01, and this allows signals for two different radicals to be separated and identified.

For the elucidation of structure using NMR spectroscopy, the appearance of the peaks (i.e. singlet, doublet, triplet, etc.) and the size of the splittings, measured by the coupling constant (J), are very important. This is the same for ESR spectroscopy, where the term coupling constant is replaced by hyperfine splitting (or hyperfine coupling or just hyperfine), which is given the symbol a. This describes the interaction of the unpaired electron with neighbouring magnetic nuclei, and the size of the splittings, which are much larger than for ^{1}H NMR spectroscopy, are reported in units of magnetic field strength (gauss G or millitesla mT; 10 G = 1 mT). For the signal to be split, we must therefore have neighbouring nuclei with a nuclear spin (i.e. $I \neq 0$), and this includes the nuclei in Table 2.1.

Table 2.1

I	Nuclei
1/2	^{1}H, ^{13}C, ^{19}F, ^{31}P
1	^{14}N, ^{2}H
3/2	^{35}Cl, ^{37}Cl, ^{79}Br, ^{81}Br
5/2	^{127}I

Each of these nuclei can cause splitting in the ESR spectrum when near to the unpaired electron. The patterns of the ESR spectra can be used to determine the number and type of magnetic nuclei adjacent to the free-radical centre, leading to the identification of the radical.

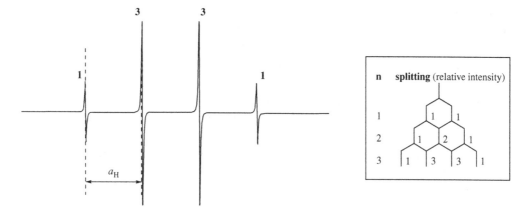

ESR spectrum of the methyl radical ($^\bullet$CH$_3$)

Fig. 2.5

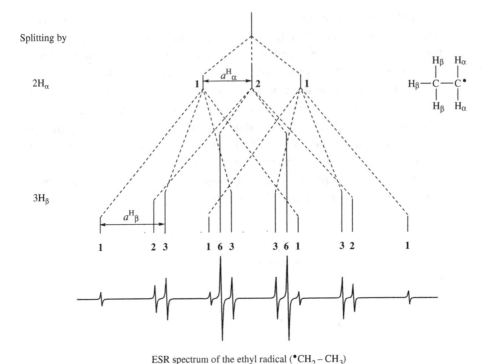

Splitting by

$2H_\alpha$

$3H_\beta$

ESR spectrum of the ethyl radical ($^\bullet CH_2 - CH_3$)

Fig. 2.6

As an example, we will consider the (first derivative) ESR spectrum of the methyl radical ($CH_3^\bullet$), shown in Fig. 2.5. The four lines (quartet) in the spectrum are derived from hyperfine splitting by the three equivalent hydrogens on the carbon to give a 1 : 3 : 3 : 1 pattern. As for ^1H NMR, we can represent this in a tree or pyramid diagram, and make use of the n + 1 rule (where n is the number of equivalent coupling nuclei; in this case, n = 3). The three hydrogen atoms, which are on the same carbon atom as the radical, are known as α-hydrogens and the splitting is recorded as a_H = 23 G or 2.3 mT.

A more complex, 12-line, ESR spectrum is observed for the ethyl radical ($CH_3—CH_2^\bullet$) (Fig. 2.6). This is derived from splitting of the signal by both the methylene (CH_2) and methyl (CH_3) hydrogens. The two α-hydrogen atoms of the methylene group produce a 1 : 2 : 1 triplet (n = 2), and each of these three lines is split further by the three β-hydrogen atoms of the methyl group into a 1 : 3 : 3 : 1 quartet (n = 3) giving a total of 12 (3 × 4) lines. The sizes of the splittings are similar: a^{CH_2} = 22.4 G and a^{CH_3} = 26.9 G. In ESR publications, these are normally quoted as hyperfine values for a^H_α and a^H_β using the nomenclature shown in Fig. 2.7.

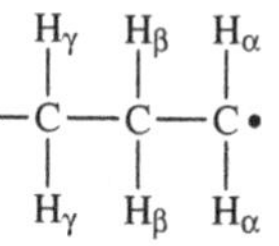

$a^H_\alpha = 21\text{--}23 \text{ G}$
$a^H_\beta = 23\text{--}30 \text{ G}$
$a^H_\gamma = < 1 \text{ G}$

Fig. 2.7

These 'typical' values for a assume that there is free rotation about the carbon–carbon bonds and that the radical is localized on the α-carbon; if the unpaired electron can be delocalized onto other atoms (attached to the α-carbon), the a^H_α splitting becomes smaller (see Section 2.4).

The size of the a value is therefore affected by both the structure of the radical and the temperature: the lower the temperature, the slower the rotation of bonds. In ^{1}H NMR spectroscopy, the size of the three-bond coupling constant (^3J) is related to the dihedral angle (ϕ) between the hydrogen atoms (in the Karplus equation) as shown in Fig. 2.8. For ESR spectroscopy, the dihedral angle between the orbital of the unpaired electron and β-C—H affects the size of the a_β splitting. Like the ^3J values, the a_β splitting is proportional to ϕ, and maximum values of a_β will be seen when $\phi = 0°$ and $180°$. For cyclobutyl and cyclohexyl radicals, the radicals are localized on the α-carbon and the rings restrict the rotation of the C—C bonds. Hence, a^H_β values can range from 39.4 G when $\phi = 0°$ to 5.3 G when $\phi = 60°$. These measurements are usually carried out at low temperature (to prevent free rotation in the molecule) and, when values of a_β are recorded over a range of temperature, this can provide information about the preferred conformations of radicals and the barriers to rotation about the C_α–C_β bond.

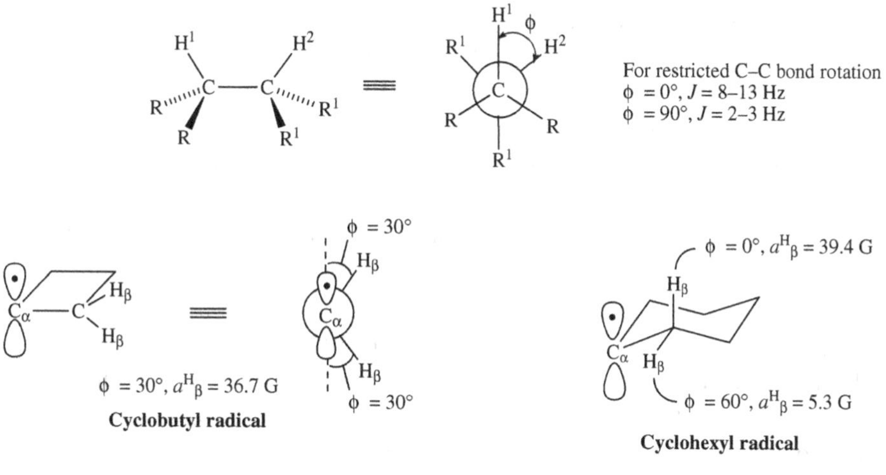

For restricted C–C bond rotation
$\phi = 0°, J = 8\text{--}13 \text{ Hz}$
$\phi = 90°, J = 2\text{--}3 \text{ Hz}$

$\phi = 30°, a^H_\beta = 36.7 \text{ G}$

Cyclobutyl radical

$\phi = 30°$

$\phi = 0°, a^H_\beta = 39.4 \text{ G}$

$\phi = 60°, a^H_\beta = 5.3 \text{ G}$

Cyclohexyl radical

Fig. 2.8

21

Computer simulations are often used to help with the identification of radicals in a mixture; trial values of the hyperfine constants are used to recreate the experimental spectrum as closely as possible. The determination of the concentrations of radicals (i.e. quantitative measurements) in a mixture is more difficult as the high reactivity of radicals means that their concentrations are often very low. In some cases, it is difficult even to detect the radicals because their concentrations are so low ($<10^{-8}$ mol dm^{-3}). To overcome this, we could use low temperatures to trap (freeze) radicals within a crystal lattice of another molecule, such as xenon. Pulse photolysis could be employed to generate a high concentration of radicals using a short burst of (high-energy) UV light, or we could employ a technique known as spin trapping.

Nitroso spin trap   Reactive radical   Nitroxide spin adduct

Fig. 2.9

As the name suggests, spin trapping involves trapping a reactive radical (R$^•$) to form a more stable and longer lived radical (R—Trap$^•$). This so-called 'spin adduct' can then be observed in the ESR spectrometer and the structure of the initial radical R$^•$ deduced. Nitroso compounds (R^{1}NO) are a common class of 'spin traps'; they react with reactive radicals to form much more stable radicals called nitroxides (Fig. 2.9). The ESR spectrum of the spin adduct derived from 2-methyl-2-nitrosopropane (MNP) and the methyl radical is shown in Fig. 2.10. The 12 lines comprise three sets of quartets derived from hyperfine splitting of the radical on the α-oxygen; the α-oxygen atom does not split the signal as $I = 0$. The β-nitrogen ($I = 1$) gives three lines ($a_N^\beta = 15.25$ G) in the ratio $1 : 1 : 1$ and these are split into quartets by the three hydrogens of the γ-methyl group ($a_H^\gamma = 11.3$ G). This splitting pattern is thus characteristic for the methyl radical adduct. Although radicals can be detected using nitroso compounds as spin adducts, the structural information obtained is generally limited. For example, the spin adduct derived from the reaction of MNP with the ethyl radical (Fig. 2.11) would exhibit nine lines in the ESR spectrum (due to splitting from the β-nitrogen and the two γ-methylene hydrogens). However, no splitting is observed from the three δ-methyl hydrogens (these are too remote from the radical centre), and so the assignment of a methyl rather than an ethyl or isopropyl group, etc. at the δ-position is not possible. Therefore, adducts derived from methyl, ethyl or isopropyl radicals, etc. cannot be distinguished. This is one reason why MNP, with a *tert*-butyl (Me₃C) side chain, is a useful spin trap—the *tert*-butyl group does not cause any splitting and the three methyl groups (at the δ-position) do not interact with the unpaired electron on oxygen.

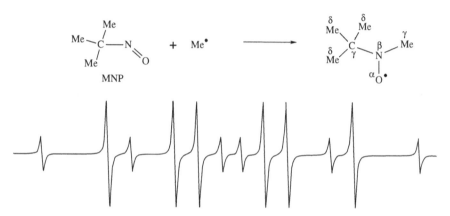

The ESR spectrum of the methyl adduct of MNP (2-methyl-2-nitrosopropane)

Fig. 2.10

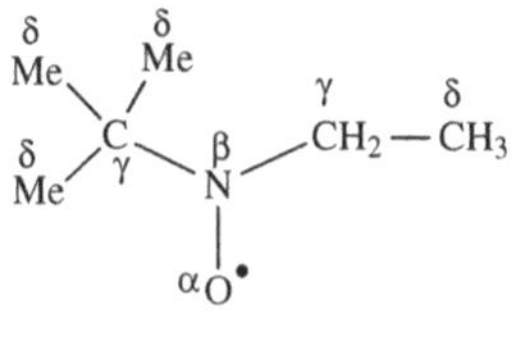

The ethyl adduct of MNP

Fig. 2.11

There is no doubt that ESR spectroscopy is a very important tool for elucidating reaction pathways. Radical intermediates can be confirmed either by direct observation or by spin trapping using an ESR spectrometer. Unfortunately, the results are not always conclusive. A minor reaction pathway could lead to a stable radical which is observed by ESR spectroscopy, but the major reaction pathway could involve a more reactive radical (which is not observed) or even ionic intermediates. In this case, radical intermediates (together with a reaction mechanism) could only be proposed for the formation of a by-product isolated in low yield.

2.2.2 Chemically induced dynamic nuclear polarization

This method involves the direct observation of radical reactions using ^{1}H NMR spectroscopy. Radical intermediates give rise to variations in the NMR signal intensities: some peaks have increased intensities (absorption or A lines) while others are negative (emission or E lines) (Fig. 2.12). CIDNP is observed for reactions involving a radical pair ($R^{1\bullet}$ and $R^{2\bullet}$), and products generated from these radical pairs (e.g. R^1R^2) can show these perturbed signal intensities, A or E, shortly after their formation. This can provide information on the type of reaction that the radical pair undergoes in solution. A detailed discussion of this phenomenon is outside the scope of this book and only a cursory explanation, based on a radical pair mechanism, is given below. (The interested reader is referred to the Further Reading section at the end of this chapter.)

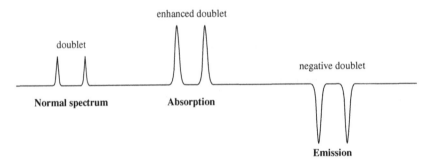

Fig. 2.12

Let us take as an example two methyl radicals (Fig. 2.13). These could either combine (couple) to give ethane or give an alternative non-radical product in the presence of other radicals, the solvent or additional reagents. Reaction with iodine, for example, would produce iodomethane (CH_3I). The ^{1}H NMR spectrum, recorded during the reaction, could show CIDNP effects for the (singlet) peaks due to ethane and CH_3I. This originates from the different electron spin states ($m_s = +1/2$ and $-1/2$). When both methyl radicals have antiparallel spins (singlet state), they can combine to form ethane, whereas radicals with parallel spins (triplet state) cannot combine as only electrons of opposite spins can 'pair'. The triplet radical pair must undergo 'spin inversion' before coupling can take place. It is found that the products (CH_3CH_3 and CH_3I) are formed from different populations of the singlet and triplet radicals. This alters the populations of the nuclear spin states in the products, and the hydrogen nuclei then absorb or emit radiation to give an equal population of spin states (called relaxation). A ^{1}H NMR spectrum recorded during this equilibration will show the enhanced or negative peaks characteristic of CIDNP. As for ESR spectroscopy, the presence of CIDNP does not prove that the main reaction involves radicals and its absence is not conclusive evidence for the absence of radical intermediates. Indeed, radical reactions have been shown to take place without observable CIDNP.

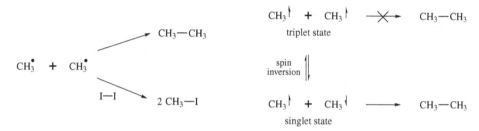

Fig. 2.13

2.3 The shape of radicals

The spectroscopic techniques highlighted above have provided considerable information on the shape of a variety of radicals. For carbon-centred radicals, $R_3C^\bullet$, the unpaired electron could be in a p orbital or an sp^3 hybrid orbital (Fig. 2.14). This would give a planar or pyramidal shape, respectively, but the shape could also be somewhere between the two. All three possibilities are known and the shape depends on the size and electronic properties (e.g. electronegativity) of the R groups. Alkyl radicals are planar or near-planar; radicals with large R groups prefer a planar arrangement because this minimizes (unfavourable) steric interactions. There are some exceptions: strained ring systems (e.g. the cyclopropyl radical) cannot adopt a planar shape, and those with heteroatoms, such as fluorine, tend to be pyramidal. Thus, whereas the *tert*-butyl and methyl radicals are planar (like the corresponding cations), the trifluoromethyl radical is pyramidal and the hydroxymethyl radical ($^\bullet CH_2OH$) is in-between. Related metal-centred radicals, $R_3M^\bullet$ (where M can be tin, silicon or germanium), are also known to be pyramidal.

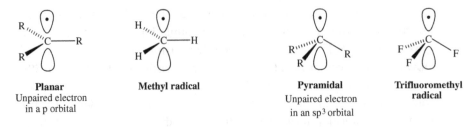

Planar
Unpaired electron
in a p orbital

Methyl radical

Pyramidal
Unpaired electron
in an sp³ orbital

**Trifluoromethyl
radical**

Fig. 2.14

2.4 Radical stability

Almost all radicals are very reactive intermediates with short lifetimes. In order to compare the stability of these intermediates, we first need to compare how easily they are made and, secondly, how quickly they react with other molecules (Fig. 2.15). The formation of radicals can be described in terms of thermodynamic stability, which is related to how easily bonds can be broken homolytically to form radicals. Once formed, not all radicals react at the same rate and they have a range of lifetimes. We will now look at each of these factors in turn.

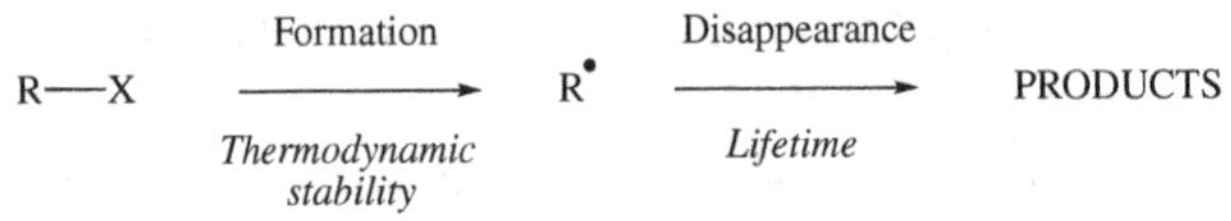

$$R\!-\!X \xrightarrow[\textit{Thermodynamic stability}]{\text{Formation}} R^{\bullet} \xrightarrow[\textit{Lifetime}]{\text{Disappearance}} \text{PRODUCTS}$$

Fig. 2.15

2.4.1 Thermodynamic stability

The energy required to cleave a bond homolytically to give radicals is called the bond dissociation energy (BDE) or bond strength (Fig. 2.16). Conversely, when two radicals combine, energy (heat) corresponding to the BDE of the newly formed bond is released. The size of the BDE depends primarily on the thermodynamic stability of the product radicals; the more stable the radicals, the lower the BDE. When reactive radicals

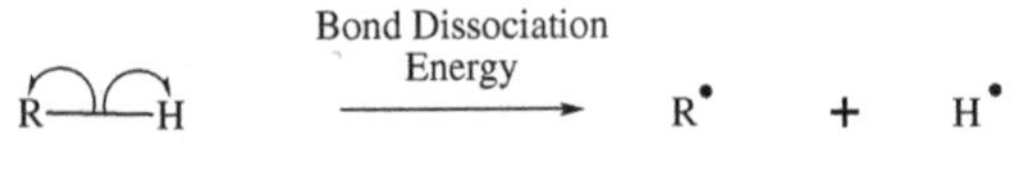

$$R\!-\!H \xrightarrow[]{\substack{\text{Bond Dissociation} \\ \text{Energy}}} R^{\bullet} \ + \ H^{\bullet}$$

Fig. 2.16

combine, more energy is released (i.e. a stronger bond is made) than that derived from the reaction of more stable radicals.

BDEs for the homolysis of specific bonds in a variety of molecules, together with the average bond energies (calculated from a range of related molecules), are shown in Tables 2.2–2.4. These are approximate values determined in the gaseous phase and are continually being revised because of improved methods of measurement. We can use these values to compare the stability of radicals in solution as well as in the gaseous phase.

Table 2.2

Bond	H—H	H—F	H—Cl	H—Br	H—I	H—OH	H—NH$_2$
BDE (kJ mol^{-1})	435	565	431	364	297	498	448

BDE, bond dissociation energy.

Table 2.3

Bond	H$_3$C—H	H$_3$C—F	H$_3$C—Cl	H$_3$C—Br	H$_3$C—I	H$_3$C—OH	H$_3$C—NH$_2$
BDE (kJ mol^{-1})	440	460	356	297	239	389	356

BDE, bond dissociation energy.

Table 2.4

Bond	R$_3$C—H	RO—H	R$_2$N—H
Average bond energy (kJ mol^{-1})	415	442	406

The strongest bonds are formed when orbitals of similar energy and size overlap. For the halogens, as we go from fluorine to iodine, the outer p orbital (which is used for bonding) becomes larger and more diffuse, leading to less efficient overlap with the 1s orbital of hydrogen. This is supported by the different bond lengths (0.92 Å for H—F and 1.60 Å for H—I). The hydrogen–iodine bond is therefore easier to break than the hydrogen–fluorine bond, and we can therefore deduce that the iodine radical (or atom) is more stable than the fluorine radical.

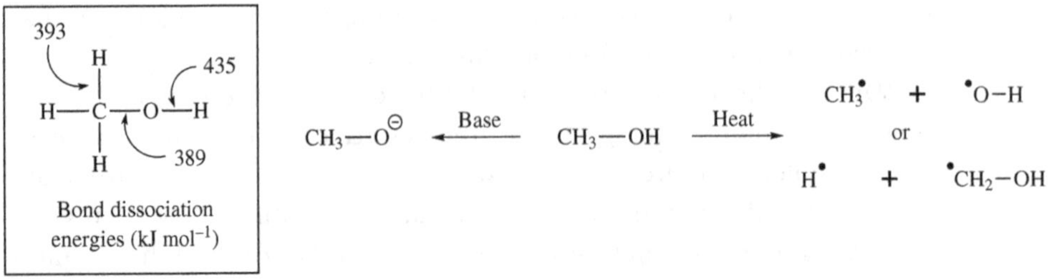

Fig. 2.17

The BDEs also show that O—H and N—H bonds are relatively strong. If we heated methanol, we would find that the O—H bond is not the first to break (Fig. 2.17). This is in contrast to the familiar heterolytic cleavage of these bonds (in the presence of bases) in solution. The anions formed on deprotonation are stabilized by the electronegative oxygen or nitrogen atoms and the solvent (solvation).

For simple alkanes, such as ethane, the carbon–hydrogen bonds are stronger than the carbon–carbon bonds (Fig. 2.18). The strength of these bonds changes with substitution (branching); the more alkyl groups on a carbon, the weaker the (remaining) C—H bond. For unsaturated hydro-carbons, with strong carbon–carbon double or triple bonds, the C—H bonds of the sp^2 or sp carbon are also much higher in energy than for alka-nes. However, the C—H BDE for a methyl group attached to an alkene, alkyne or benzene ring has been shown to be much smaller than for alka-nes. We can understand why these BDEs vary by considering the factors known to contribute to the thermodynamic stability of the radicals which are formed.

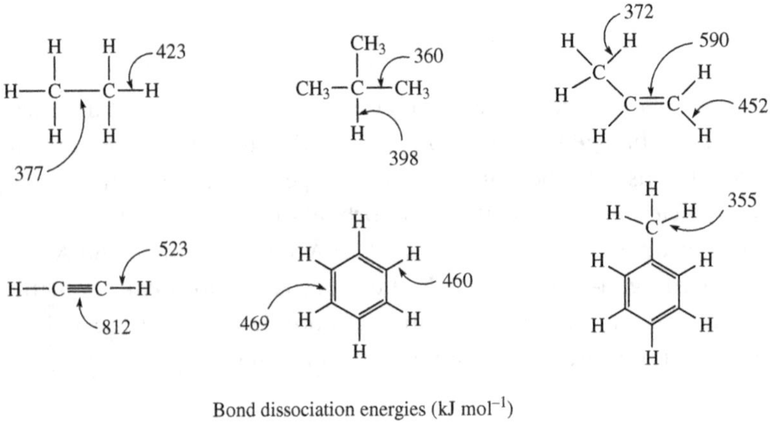

Bond dissociation energies (kJ mol^{-1})

Fig. 2.18

2.4.1.1 Hyperconjugation

Although electronically neutral, radicals are electron deficient. Carbon-centred radicals have only seven electrons in the outer shell and are one electron short of the energetically favoured octet (eight). We might therefore expect that, like carbocations (R_3C^+), radicals would be stabilized by electron-donating substituents pushing electron density towards the electron-deficient carbon. This is indeed the case, and alkyl radicals follow the same order of stability as carbocations (Fig. 2.19). A tertiary radical ($R_3C^•$) is therefore more stable than a methyl radical ($CH_3^•$) because of the three electron-releasing alkyl groups that exert a positive inductive effect (+I). As a tertiary radical is more stable, it is easier to prepare, and therefore the C—H bond of a tertiary alkane is much weaker than that for methane (i.e. 400 vs. 440 kJ mol^{-1}). We should also note the contribution due to steric crowding; the formation of a tertiary radical leads to the greatest release of strain (Fig. 2.20). The more bulky (R) groups present in the hydrocarbon, the more strained the molecule will be, and hence the greater the relief of strain on going from an sp^3 (tetrahedral) to an sp^2 (trigonal planar) carbon, in which the R groups are further apart.

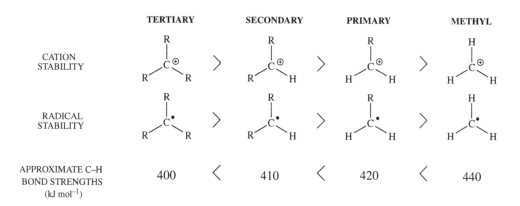

Fig. 2.19

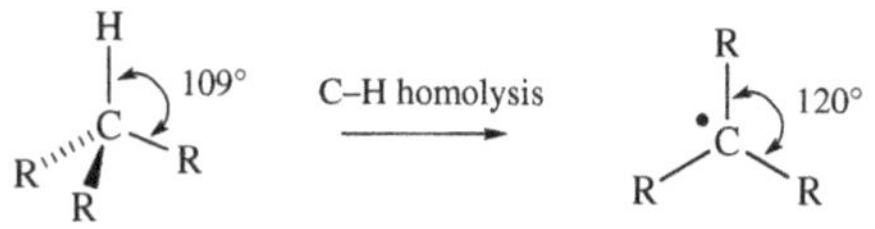

Fig. 2.20

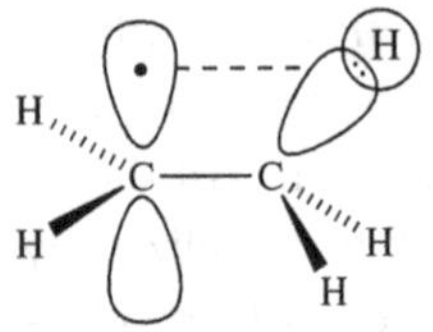

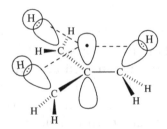

Hyperconjugation in the
ethyl radical (CH₃CH₂•)

Hyperconjugation in the
tert-butyl radical (Me₃C•)

Fig. 2.21

The stabilization of radicals by alkyl groups can also be accounted for by a phenomenon known as hyperconjugation. This describes the interaction of the p orbital of the radical with a pair of bonding electrons in a neighbouring σ bond (Fig. 2.21). Electrons in the filled (C_{sp^3}—H_{1s}) σ bond can be donated to the partly filled p orbital of the radical. The resulting delocalization, like the delocalization of electrons in benzene, has a stabilizing effect. As the hydrogen atoms are successively replaced by alkyl groups, the number of σ bonds which can interact with the p orbital increases, leading to more stable radicals. Thus, whereas the (primary) ethyl radical interacts with three adjacent C—H bonds, the more stable tert-butyl radical can interact with nine.

Hyperconjugation is responsible for the β-hyperfine splittings (a^H_β) observed in the ESR spectrum of the ethyl radical (see Section 2.2.1). The electron density is delocalized from the odd electron to the methyl hydrogens and this can be represented using one-electron arrows (Fig. 2.22). The extent of hyperconjugation depends upon the dihedral angle (ϕ) between the C—H bond and the orbital of the unpaired electron, and this can be determined from the size of the a^H_β splittings in the ESR spectrum.

Fig. 2.22

2.4.1.2 Resonance

Resonance (or mesomerism) is similar to hyperconjugation in that it leads to electron delocalization and radical stabilization, and hence lower BDEs. Whereas hyperconjugation involves overlap of the p orbital with a σ bond, resonance normally describes the interaction of the p orbital with a π bond or lone pair. Stronger orbital interactions are observed in resonance than in hyperconjugation, leading to even more stable radicals.

As shown in Fig. 2.18, the allylic and benzylic C—H bonds in propene and toluene (methylbenzene), respectively, are much weaker than the C—H bonds in ethane (i.e. 372 and 355 vs. 423 kJ mol^{-1}). These C—H bonds, which are adjacent to an alkene double bond or a benzene ring, are weaker because of the stability of the resulting allyl or benzyl radicals. The single p electron can combine (or conjugate) with the π electrons, and maximum (p–π) overlap results when the carbon atom bearing the radical is planar. As the unpaired electron is in a π orbital, this is known as a π-type radical. For the allyl radical, we can represent this using one-electron curly arrows to give two degenerate (equivalent energy) resonance hybrids or canonical structures as shown in Fig. 2.23. As a rule of thumb, the more resonance hybrids that can be drawn, the greater the stability of the radical. For the benzyl radical, we can draw four resonance hybrids (two of which are equivalent). These, however, do not make equal contributions and the resonance hybrid which retains the aromatic benzene ring is the most important.

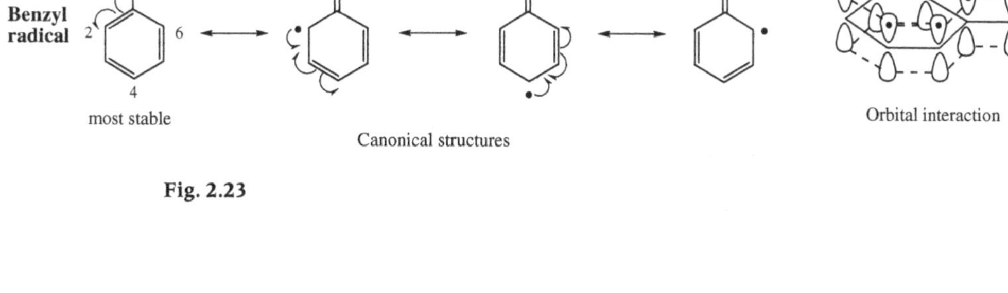

Fig. 2.23

These resonance forms are consistent with ESR measurements. The electron density at a particular atom is proportional to the size of splitting of the ESR peaks, and the greater the electron density, the greater the coupling constant (a). The allyl radical has been shown to have most of its electron density at C-1 and C-3, whereas the electron distribution in the benzene ring of the benzyl radical has been shown to be higher for the 2-, 4- and 6-positions.

Radicals centred on atoms other than carbon can also be stabilized by resonance. For the phenoxyl radical, the unpaired electron in the oxygen p orbital can combine with the π orbital of the benzene ring (Fig. 2.24). This radical delocalization explains why the O—H bond in phenol (360 kJ mol^{-1}) is much weaker than that in an aliphatic alcohol (e.g. 435 kJ mol^{-1} for methanol).

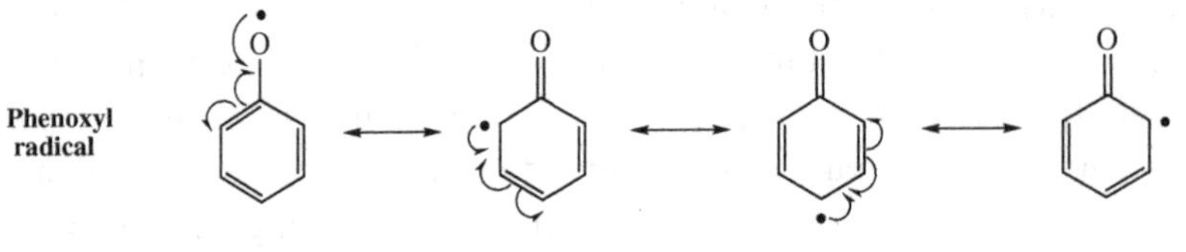

Phenoxyl
radical

Fig. 2.24

Other π bonds can interact with the p orbital of the radical leading to stabilization (Fig. 2.25). Radicals formed at the α-position of nitriles or carbonyl compounds (aldehydes, ketones, esters, carboxylic acids, etc.) are stabilized by conjugation with the π bond of the C≡N or C=O group. The electron density is spread between the α-carbon and the heteroatom; the greater the stabilization, the weaker the α-C—H bond (Fig. 2.26).

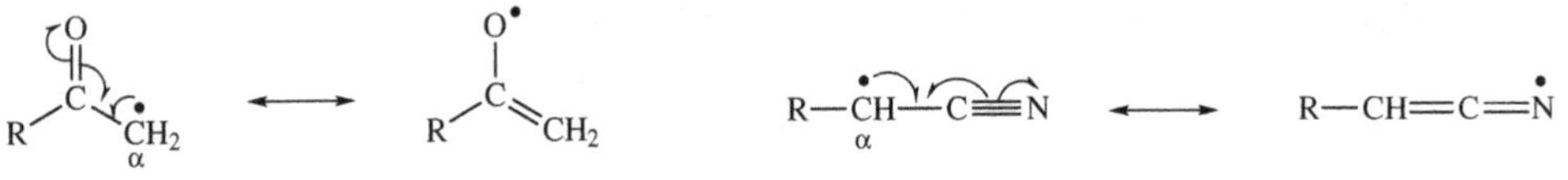

Fig. 2.25

H—CH$_2$—X   ⟶   H•   +   •CH$_2$—X

Substituent (X)	CH$_3$	CO$_2$CH$_3$	CO$_2$H	COCH$_3$	CHO	CN
C–H bond strength (kJ mol^{-1})	423	414	406	385	385	360

weaker bond ⟶

Fig. 2.26

We can draw a parallel here with carbanion chemistry. Carbonyl compounds, for example, can be deprotonated at the (acidic) α-position because the resulting enolate is stabilized by resonance (Fig. 2.27). In the corresponding radical, the unpaired electron is also delocalized onto the electronegative oxygen atom.

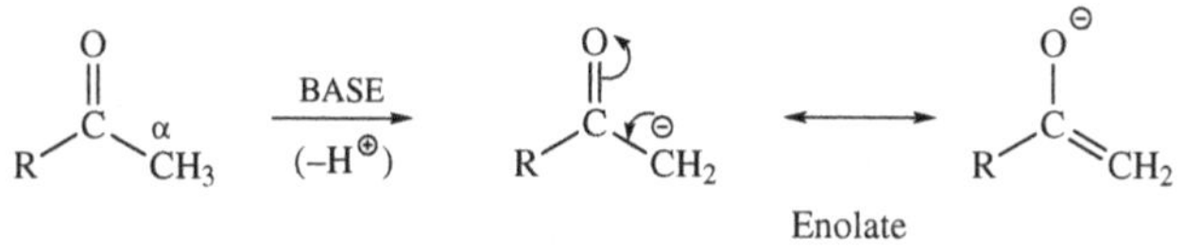

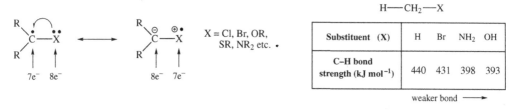

Fig. 2.27

These interactions involve mesomerically electron-withdrawing substituents, known as −M groups, but we can also stabilize radicals using mesomerically electron-donating groups (+M). The orbital of the radical can overlap with non-bonded (n) electrons present on heteroatoms, leading to 'three-electron bonding'—the single electron of the radical interacts with two electrons on the heteroatom (Fig. 2.28). Heteroatoms which can stabilize adjacent radicals include oxygen, nitrogen, sulfur and, to a lesser extent, the halogens (Br, Cl, etc.). Conjugation of the radical with the lone pair produces a separation of charge as an electron is donated from the heteroatom. For very electronegative atoms (e.g. the halogens), this is unfavourable and, consequently, we observe little stabilization of the radical; hence, the C—H bond strengths are similar in methane and bromomethane.

$$H—CH_2—X$$

Substituent (X)	H	Br	NH$_2$	OH
C–H bond strength (kJ mol^{-1})	440	431	398	393

weaker bond ⟶

X = Cl, Br, OR, SR, NR$_2$ etc. ·

7e$^-$ 8e$^-$ 8e$^-$ 7e$^-$

Fig. 2.28

The lone pair on the oxygen of a carbonyl group can stabilize an acyl radical in a similar manner (Fig. 2.29). This explains the low BDE for aldehydic hydrogens, and in ethanal this is the weakest C—H bond.

Acyl radical

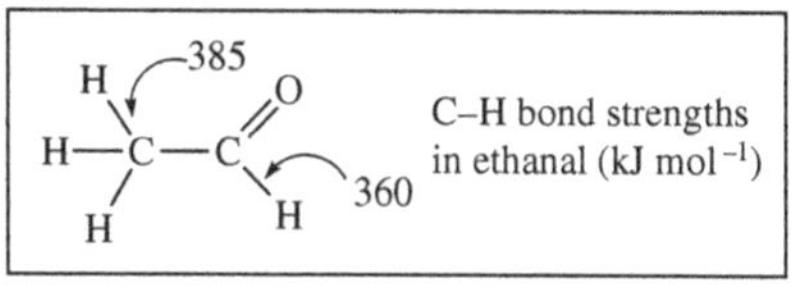

C–H bond strengths in ethanal (kJ mol^{-1})

−385

360

Fig. 2.29

Radicals centred on atoms other than carbon can be stabilized in a similar way. The O—H bond of hydroxylamines, $R_2NO—H$, is weaker than that of alcohols (i.e. 310 vs. 435 kJ mol^{-1}) because the nitroxide radical is stabilized by conjugation (with the nitrogen atom), as shown by the two resonance structures in Fig. 2.30. Nitroxides have been exploited in ESR spin trapping experiments: reaction of short-lived carbon radicals with a nitrone, such as 5,5-dimethylpyrroline-*N*-oxide (DMPO), produces spin adduct nitroxides which can be observed by ESR spectroscopy (cf. nitroso spin traps; see Section 2.2.1).

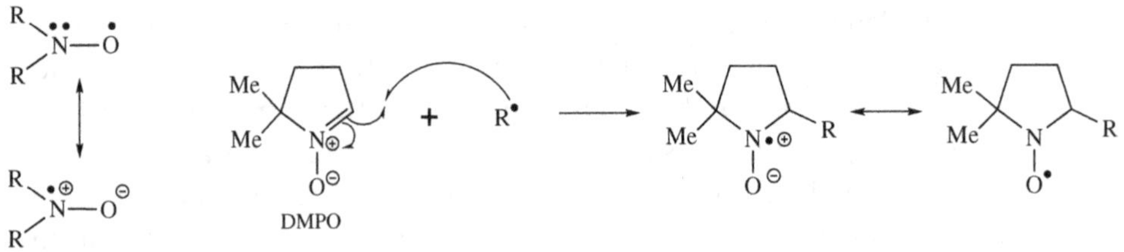

Fig. 2.30

When we introduce more than one conjugating substituent at the radical centre, the C—H bond does not necessarily become weaker. A dialdehyde has a stronger (methylene) C—H bond than an aldehyde, whereas a dimethyl acetal has a slightly weaker C—H bond than a methyl ether (Table 2.5). Pairs of the same substituents can therefore act in a conflicting manner even though more resonance structures can be drawn. This is surprising as we might expect the mesomeric effects of two carbonyl groups to stabilize a radical more effectively than one carbonyl substituent (as in carbanion chemistry). However, we also need to consider the inductive effect of the carbonyl group which, because it is electron withdrawing (or –I), will *destabilize* the (electron-deficient) radical. We therefore have a competition between mesomeric stabilization and inductive destabilization. Although mesomeric effects are usually stronger than inductive effects, it appears that, with the dicarbonyl radical, the two –I effects exert a greater influence than the two –M effects of the two groups.

Table 2.5 Approximate bond dissociation energies (kJ mol^{-1}).

H—CH$_2$CHO	385	H—CH$_2$OMe	389	H—CH$_2$CO$_2$H	406
H—CH(CHO)$_2$	414	H—CH(OMe)$_2$	381	H—CH$_2$NH$_2$	398
				H—CH(NH$_2$)CO$_2$H	318

The influence of disubstitution is particularly pronounced for molecules containing electron-withdrawing ('capto') and electron-donating ('dative') groups. The energy of the α-C—H bond of glycine is only 318 kJ mol^{-1} and the radical is stabilized by both the amine (+M) and acid (–M) groups (Table 2.5). These radicals can be even more stable than would be predicted from the combined effects of the two substituents. It has been suggested that electron 'push (+M) and pull (–M) groups' work in unison to give a so-called captodative stabilization which we can represent by five resonance structures; this is shown for the glycine radical in Fig. 2.31.

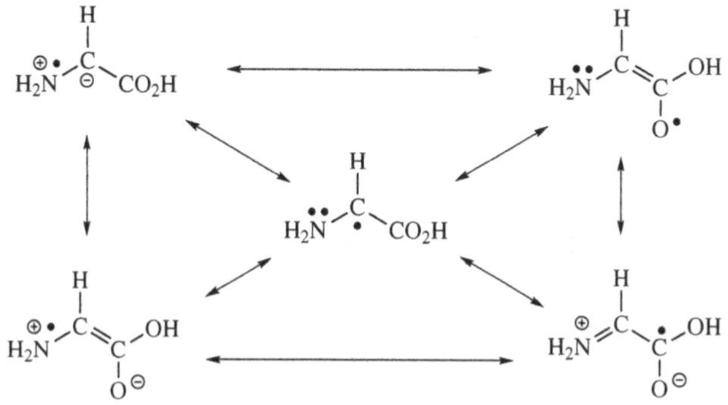

Fig. 2.31

2.4.1.3 Hybridization

We saw earlier that alkene, alkyne and benzene C—H bonds are much stronger than those of alkanes (Fig. 2.18). This is a consequence of the different hybridization of the bonds: sp^3 for alkanes, sp^2 for alkenes/benzenes and sp for alkynes. The 's' character of the C—H bond therefore increases as we move from a single to a triple carbon–carbon bond. Homolysis of the C—H σ bond in unsaturated hydrocarbons gives rise to radicals in which the orbital of the unpaired electron has a high degree of 's' character (sp^2 or sp), and so lies at right angles to the π orbitals of the C=C and C≡C bonds (Fig. 2.32). The unpaired electron remains in the σ orbital of the carbon atom and these radicals are known as σ-type radicals. As σ-type radicals are at 90° to π orbitals, no resonance delocalization (or orbital overlap) can occur, and this is supported by ESR studies on the phenyl radical (trapped in a matrix at low temperature). Measurements have shown that the unpaired electron density is chiefly located on the carbon at which the C—H bond was broken (cf. the electron spin density observed at positions 2, 4 and 6 of the benzene ring for the π-type benzyl radical). The resultant aryl, vinyl and alkynyl ($RC\equiv C^{\bullet}$) radicals are very unstable, and high energies (>450 kJ mol^{-1}) are required for cleavage of the C—H bonds.

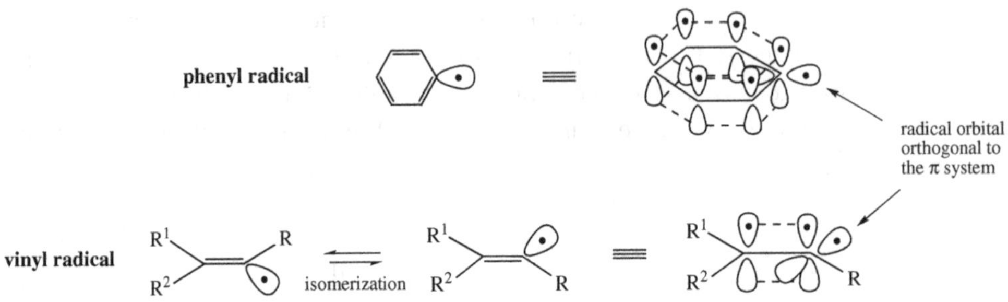

Fig. 2.32

Interestingly, as the unpaired electron in σ-type radicals is not in the same plane as the π electrons, vinyl radicals can have geometric isomers. Unlike alkenes, however, these isomers (designated Z and E or cis and trans) are found to undergo rapid interconversion, the barrier to inversion of the carbon atom bearing the radical being very low (typically 8 kJ mol^{-1}).

2.4.2 Radical lifetime

Whereas the thermodynamic stability of radicals is related to electronic effects, the lifetime is generally determined by steric factors. The larger the substituents around a radical centre, the more stable the radical will be, and the greater its lifetime. Radical lifetimes are generally reported as half-lives ($t_{1/2}$) at a given concentration, and represent the time taken for the concentration of radicals to fall to half the initial value. The longer the half-life, the more stable (or less reactive) the radical. In solution, very reactive alkyl radicals will react with themselves (dimerize) every time they collide in extremely fast 'diffusion-controlled reactions'. Very short half-lives are recorded and, for example, the reactive methyl radical has a half-life of only 0.2×10^{-3} s at a concentration as dilute as 10^{-6} mol dm^{-3}! At the same concentration, the much larger triisopropylmethyl radical, ($Me_2CH)_3C^{\bullet}$, has a half-life of 21 h. Steric factors, as well as hyperconjugation, are responsible for the increased stability of tertiary radicals, and this is particularly important for congested radicals such as the triisopropylmethyl radical.

Steric effects are also important for stabilization of the unreactive triphenylmethyl radical (Fig. 2.33). At first sight, resonance stabilization of the
unpaired electron by all three benzene rings would appear to be responsible. However, the three benzene rings would have to be in the same plane
for maximum overlap of the p orbital with the aromatic π orbitals.
Spectroscopic and X-ray crystallographic measurements have shown that
this is not the case. The benzene rings are twisted like the blades of a propeller. Therefore, delocalization of the unpaired electron in Ph$_3$C$^\bullet$ (2) is restricted and it is not substantially greater than that observed for the benzyl
radical. However, the triphenylmethyl radical has a longer half-life and is
more reluctant to undergo dimerization than the benzyl radical. This is
due to a steric effect as the three (angled) benzene rings effectively shield the
radical from dimerization. The shielding is so effective that, when dimerization does occur, one of the radicals that reacts is on a more exposed
carbon atom of the benzene ring (to give compound (4)).

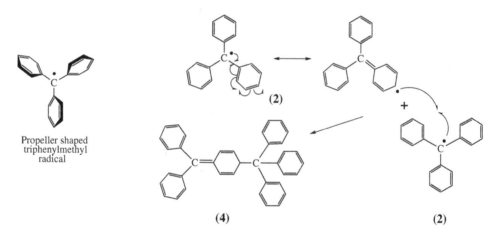

Propeller shaped
triphenylmethyl
radical

(2)

+

(4)

(2)

Fig. 2.33

Some radicals, including the triisopropylmethyl and triphenylmethyl
radicals, which are stabilized by both thermodynamic and kinetic factors, can therefore have half-lives of greater than 10^{-3} s. Relatively high
concentrations of these radicals can be maintained in solution (as they
react relatively slowly with the solvent or each other) and, as a result, they
can be easily studied using ESR spectroscopy. These radicals with long
lifetimes are said to be 'persistent radicals'. Some of them, including 2,2,6,6-
tetramethyl-1-piperidinyloxyl (TEMPO), Fremy's salt and 1,1-diphenyl-
2-picrylhydrazyl, are so stable that they can be isolated and stored for
many months without decomposing (Fig. 2.34).

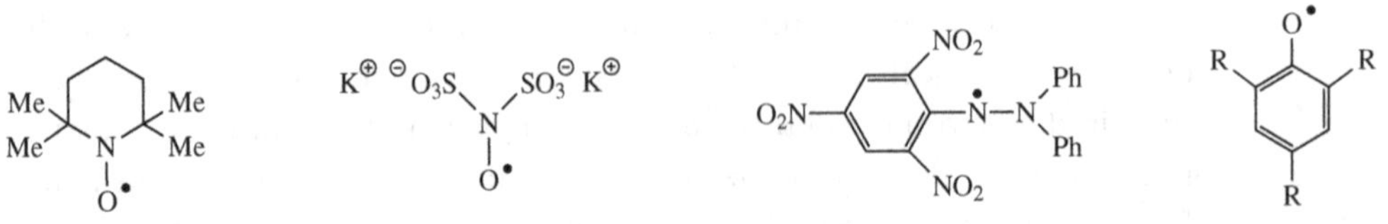

TEMPO	Fremy's salt	1,1-Diphenyl-2-picrylhydrazyl, DPPH	2,4,6-Substituted phenoxyl radical

Fig. 2.34

Of particular industrial and biological importance are persistent phenoxyl radicals. These are derived from the homolysis of the weak O—H bond of substituted phenols; the resultant radicals (of the type shown in Fig. 2.34) are stabilized by both resonance and steric hindrance. This has been exploited in the food and automobile industries where phenols, including butylated hydroxyanisole (BHA) and butylated hydroxytoluene (BHT), are used to prevent the oxidative decomposition of fats and oils (see Section 1.4). The phenols readily donate a hydrogen atom (to form the stable phenoxyl radicals) and so quench the reactive radicals involved in the oxidation process. Related antioxidants are also present in nature, and substituted phenols, such as vitamin E, are able to protect the body from harmful oxidative radicals in a similar manner (see Section 1.3).

2.4.3 A frontier molecular orbital approach

We have seen that radicals can be stabilized by both electron donors and acceptors, and this can be explained in terms of conjugation. More recently, a frontier molecular orbital approach has been developed to show the orbital interactions responsible for this stabilization. This approach considers interactions of the highest occupied (HOMO) or lowest unoccupied (LUMO) molecular orbitals, otherwise known as frontier orbitals.

A molecular orbital approach can be used to explain the stability of the allyl radical, $CH_2{=}CH{-}CH_2{\cdot}$. As discussed earlier in Section 2.4.1.2, this stability results from interaction of the p orbital (of the radical) with two p orbitals of the C=C π bond. The three (carbon) p orbitals are of equal energy and these combine to produce three molecular orbitals of different energy (Fig. 2.35). When all three atomic orbitals are in the same phase, they combine to give the lowest energy molecular orbital (π_1), which is called the bonding orbital. The next orbital has no electron density (i.e. a node) at the middle carbon, and this is called the non-bonding orbital (π_2) because there are no orbital interactions (it has the same energy as the p orbitals). The highest energy molecular orbital (with two nodes) results from the interaction of three out-of-phase atomic orbitals, and this is known as the antibonding orbital (π_3). Each molecular orbital can only accommodate two electrons. As there are three electrons in the allyl radical, two populate the bonding orbital, and the third, the non-bonding orbital. The energy of this system is lower (i.e. more favourable) than that for three non-interacting p orbitals because the low-energy bonding orbital (π_1) is filled (and the high-energy antibonding orbital, π_3, is empty). The π_2 orbital is called the singly occupied molecular orbital (SOMO), and the single electron in this orbital is located at either end of the carbon chain (there is a node at the central carbon). This is consistent with the resonance structures and the electron distribution measured using ESR spectroscopy.

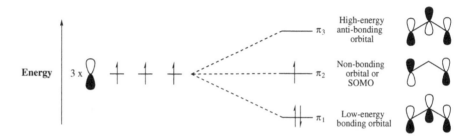

The molecular orbitals of the allyl radical ($CH_2=CH-CH_2\bullet \leftrightarrow \bullet CH_2-CH=CH_2$)

Fig. 2.35

We can explain the orbital interactions leading to radical stabilization by heteroatom substituents in a similar way. For a radical adjacent to an electron-rich substituent (e.g. $R_2C\bullet{-}NR_2$), we need to consider the interaction of the p orbital (or SOMO of the radical) with the lone pair of the electron-donating group (Fig. 2.36a). The lone pair of electrons fills a non-bonding (n) orbital (the HOMO) and this is of similar energy to the SOMO of the radical. As they are close in energy they can interact, and this produces two new molecular orbitals. Two of the three electrons reside in the lower energy molecular orbital. The remaining electron is in the new higher energy SOMO. As two of the three electrons have decreased in energy, this leads to an overall stabilization of the system.

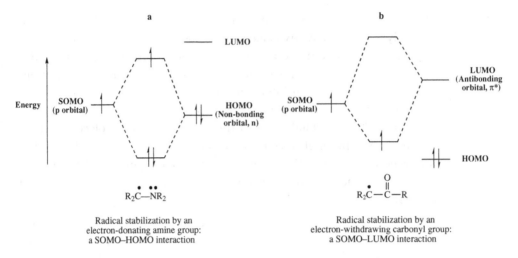

Fig. 2.36

For electron-withdrawing (–M) groups, such as carbonyls, we first need to consider which orbital of the C=O group will interact with the p orbital of the radical [in $R_2C^{\bullet}$—C(O)R] (Fig. 2.36b). This will depend on their energies, as only orbitals of similar energy can interact significantly. For carbonyls, the bonding (HOMO) and antibonding (LUMO) orbitals of the π system are the highest in energy and it is the unfilled antibonding (or π*) orbital which is closest in energy to the p orbital (or SOMO) of the radical. Two new molecular orbitals are formed, and the single electron enters a new, lower energy SOMO and the radical is stabilized.

In contrast to ions (carbocations and carbanions), radicals can therefore be stabilized by *both* electron-donating and electron-withdrawing groups. This is because the carbon p orbital, or SOMO, of the radical can interact with either filled (HOMO) or vacant (LUMO) orbitals. Radicals adjacent to electron-donating groups interact with a filled orbital to give a high-energy SOMO, while radicals next to electron-withdrawing groups interact with an unfilled orbital to give a low-energy SOMO. As radicals with different substituents have different SOMO energies, they will react differently (see Section 4.7.5). Radicals with a low-energy SOMO generally interact with a filled orbital, to accept an electron, and show electrophilic properties. Radicals with a high-energy SOMO generally interact with an unfilled orbital, to donate an electron, and show nucleophilic properties. We can therefore think of radicals as having some polar character which is determined by the substituents: nucleophilic radicals have electron-donating (+I, +M) substituents, while electrophilic radicals have electron-withdrawing (–I, –M) substituents (Fig. 2.37).

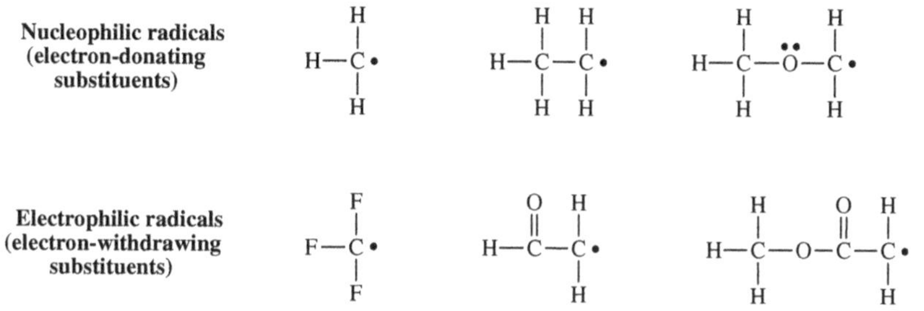
Nucleophilic radicals
(electron-donating
substituents)
Electrophilic radicals
(electron-withdrawing
substituents)

Fig. 2.37

2.5 Radical reactions: chain vs. non-chain mechanisms

We have seen that radicals are prepared by the homolytic cleavage of bonds. The resultant radicals can react with other molecules, such as alkenes (in polymerization), or, alternatively, combine with themselves or with other radicals. We can describe these very common processes by either a chain or non-chain mechanism involving a series of elementary reactions. Both mechanisms begin with radical formation in an initiation process (Fig. 2.38). The radicals ($R^\bullet$) can react with non-radical molecules (M) and/or undergo rearrangements to produce new radicals, $R^{1\bullet}$ and $R^{2\bullet}$, in propagation steps. A propagation step must have one radical precursor and one radical product. As one radical is formed at the expense of another, and the initial radical ($R^\bullet$) is regenerated, a chain reaction occurs and this can be represented in a cycle. In competition with chain propagation reactions are termination reactions, such as dimerization, which lead to the disappearance of radicals. This is where, for example, two radicals combine to give non-radical products; as no radical is regenerated, this leads to a non-chain reaction.

The next two chapters discuss the elementary steps involved in radical chain reactions. Chapter 3 describes initiation reactions which lead to the formation of radicals, whilst Chapter 4 shows how radicals react in chain propagation and termination reactions. This is illustrated by, for example, probably the first reaction encountered in general organic chemistry textbooks—the radical chain reaction, which occurs when otherwise notoriously unreactive alkanes react with chlorine gas (on heating or irradiation with UV light) to produce chlorinated hydrocarbons.

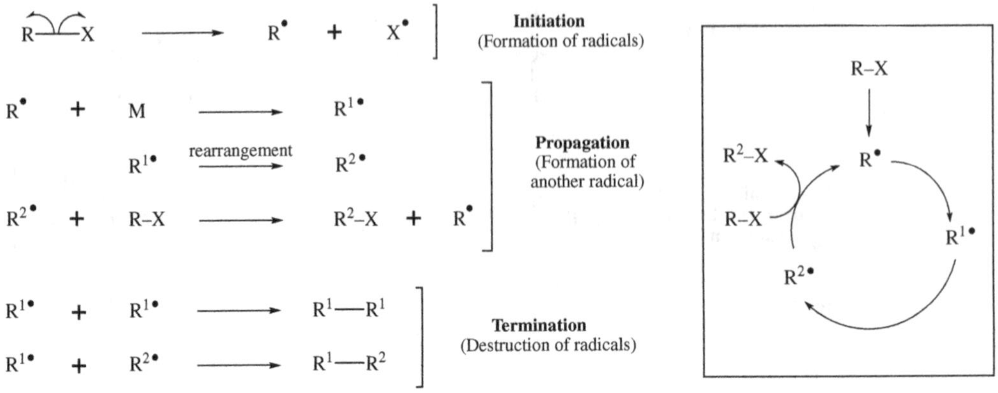

A Chain Reaction – Linear and Cyclic Representations

Fig. 2.38

2.6 Summary

Radicals can be detected and their structure deduced using a variety of spectroscopic methods, including UV spectroscopy, CIDNP and, most importantly, ESR spectroscopy. The use of these techniques has helped to show that not all radicals are the same. They have different stabilities and electronic 'characters'. Radicals can be stabilized by both electronic (orbital interactions) and steric (size of substituents) factors. Both electron-donating and electron-withdrawing substituents can stabilize radicals and, when these substituents are large, radicals can result that are long lived or persistent. These stable radicals have found application as radical inhibitors (e.g. antioxidants). Although radicals are neutral, the substituents can impart a polar character; radicals with electron-donating substituents are nucleophilic, while radicals with electron-withdrawing substituents are electrophilic. This polarity can be important for radical reactions. Radical reactions can proceed via either a chain or non-chain mechanism which involves three elementary steps. The initiation step produces the radicals, the propagation step involves radical–molecule or rearrangement reactions, while the termination step involves radical–radical reactions.

Further reading

Brocks, J.J., Beckhaus, H.-D., Beckwith, A.L.J. & Rüchardt, C. (1998) Estimation of bond dissociation energies and radical stabilization energies by ESR spectroscopy. *Journal of Organic Chemistry*, **63**, 1935–1943.

Bunce, N.J. (1987) Introduction to the interpretation of electron spin resonance spectra of organic radicals. *Journal of Chemical Education*, **64**, 907–914.

Eberson, L. (1998) Spin trapping and electron transfer. *Advances in Physical Organic Chemistry*, **31**, 91–141.

Fischer, H. & Paul, H. (1987) Rate constants for some prototype radical reactions in liquids by kinetic electron spin resonance. *Accounts of Chemical Research*, **20**, 200–206.

Griller, D. & Ingold, K.U. (1976) Persistent carbon-centred radicals. *Accounts of Chemical Research*, **9**, 13–19.

Kaptein, R. (1975) Chemically induced dynamic nuclear polarization. Theory and applications in mechanistic chemistry. *Advances in Free-Radical Chemistry*, **5**, 319–380.

Leroy, G., Sana, M., Wilante, C. & Nemba, R.M. (1989) Bond-dissociation energies of organic compounds. A tentative rationalization based on the concept of stabilization energy. *Journal of Molecular Structure*, **198**, 159–173.

Norman, R.O.C. (1970) Structures of organic radicals. *Chemistry in Britain*, **6**, 66–77.

Viehe, H.G., Janousek, Z., Merényi, R. & Stella, L. (1985) The captodative effect. *Accounts of Chemical Research*, **18**, 148–24.

Ward, H.R. (1972) Chemically induced dynamic nuclear polarization (CIDNP). I. The phenomenon, examples, and applications. *Accounts of Chemical Research*, **5**, 18–24.

CHAPTER 3

Radical Initiation

3.1 Introduction

For radical initiation (or generation), we require molecules with weak covalent bonds, so that homolysis of the bonds can occur under mild reaction conditions. The energy required to cleave the bond can be provided by heating (thermolysis), by ultraviolet (UV) light (photolysis) or by X-rays (radiolysis). Alternatively, radicals can be produced from redox reactions (oxidation or reduction). Thus, if the precursor is reacted with an alkali metal or a transition metal salt, or is treated electrochemically, electron transfer can occur and this can also lead to the formation of a radical.

3.2 Thermolysis

If we heat organic molecules to high temperatures, we would expect that a number of bonds would break, leading to a multitude of different radicals. This process has been utilized in industry, and long-chain alkanes are 'cracked' at temperatures around 600°C. At these high temperatures, however, excessive radical generation can be hazardous and explosions can occur. For predictable and 'clean' radical generation, we require molecules containing a relatively weak bond which can be selectively cleaved at low temperature. Molecules with a bond energy of around $125-165 \text{ kJ mol}^{-1}$ are therefore commonly used as radical precursors because they can undergo homolysis in solution at temperatures below approximately 150°C. As carbon–hydrogen and carbon–carbon bonds are normally too strong to be broken under these conditions, the functional groups that are cleaved generally contain single heteroatom–heteroatom, heteroatom–carbon or metal–metal bonds.

3.2.1 Peroxides and diacylperoxides

The single O—O bond of peroxides has a bond dissociation energy of approximately 150 kJ mol^{-1}. When these molecules are heated to between 50 and 150°C, homolysis produces two alkoxyl radicals ($RO^\bullet$) (Fig. 3.1). The O—O bond strength depends on the substituents (R), as shown by the comparison of di-*tert*-butylperoxide (^tBuO—O^tBu) with *tert*-butylhy-droperoxide (^tBuO—OH). The weaker O—O bond in ^tBuO—O^tBu could be due to: (i) a steric effect (with interaction of the two bulky *tert*-butyl groups weakening the bond); (ii) an electronic effect [as the *tert*-butoxyl radicals (with a +I group) are more stable than the hydroxyl radical]; or (iii) a combination of both. The bond energies are also reflected in the rates of unimolecular decomposition of these molecules. Experimental data show that decomposition of the peroxide (with a rate constant of 1.4×10^{-8} s^{-1}) is more than 700 times faster than that of the hydroperoxide (1.0×10^{-5} s^{-1}), even when the hydroperoxide is heated at a higher temperature (150°C vs. 70°C)! This argument can be extended further, as hydrogen peroxide (HO—OH) has a stronger O—O bond (213 kJ mol^{-1}) because homolysis generates two highly unstable hydroxyl radicals.

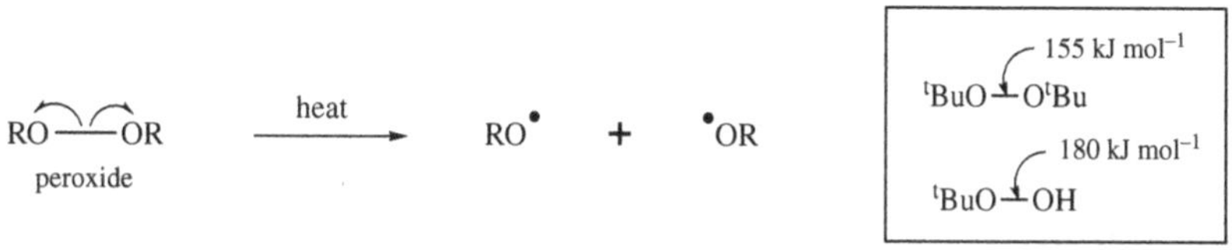

Fig. 3.1

Diacylperoxides, like peroxides, undergo homolysis of the weak O—O bond on heating (Fig. 3.2). However, these molecules have lower bond energies (≈ 125 kJ mol^{-1} for the O—O bond) than most peroxides and so are much more reactive. This is because the two acyloxyl radicals (RCO_2•) are stabilized by resonance; the electron is delocalized over both oxygens. The substituents (R) have little effect on the O—O bond strength (or rate of decomposition) and, of these initiators, benzoyl peroxide (where R = Ph) and lauroyl peroxide [where R = $CH_3(CH_2)_{10}$] are two of the most commonly employed.

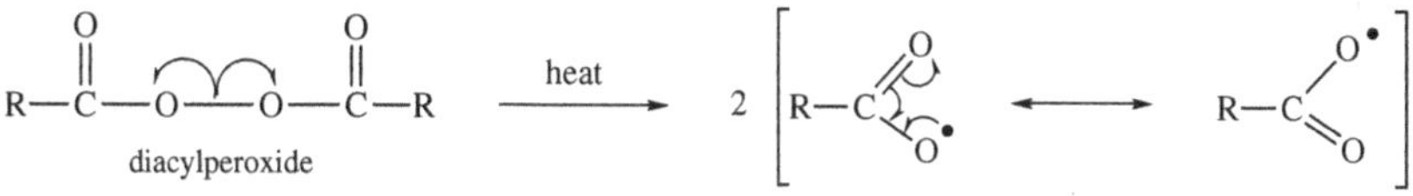

Fig. 3.2

3.2.2 Azo compounds

The thermolysis of azo compounds (R—N=N—R) leads to the cleavage of two carbon–nitrogen bonds and the formation of the stable dinitrogen molecule with a very strong N≡N triple bond (945 kJ mol^{-1}). This very high bond energy explains why N_2 is one of the best leaving groups in organic chemistry. The rate of decomposition is determined by the substituents (R); substituents which can stabilize the product radicals (R$^{\bullet}$) lead to faster cleavage of the C—N bonds. The most important azo initiator is azobisisobutyronitrile (AIBN), which is easily decomposed because of the stability of the two 2-cyano-2-propyl radicals, $^{\bullet}C(CN)Me_2$ (Fig. 3.3). These tertiary radicals are stabilized by resonance—the p orbital (of the unpaired electron) can interact with the π orbital of the nitrile triple bond and the electron-donating methyl groups.

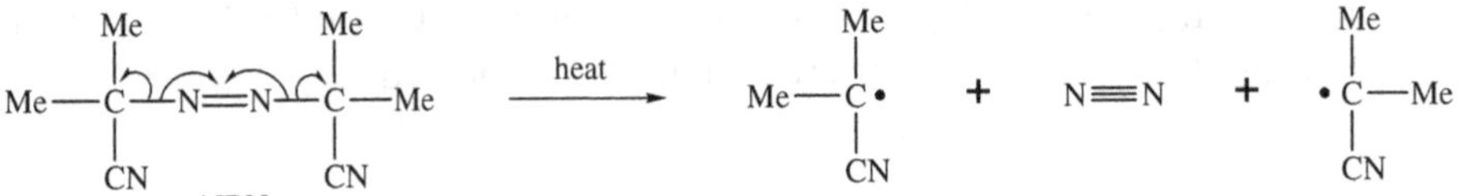

Fig. 3.3

3.3 Photolysis

If we shine visible or UV light onto an organic molecule, the light energy may be absorbed leading to the homolysis of a bond. Bonds are weaker in the 'excited state', as low-energy bonding or non-bonding electrons are promoted to a higher energy antibonding orbital (σ→σ*, n→σ*, n→π* or π→π*). The energy of light is inversely proportional to the wavelength (λ) (Fig. 3.4). Visible light, or more commonly UV light (which has a shorter wavelength and hence a higher energy), can be used to break a variety of bonds, provided that the molecule is able to absorb the light. Therefore, even though visible light carries enough energy to break a weak covalent bond, the homolysis will usually only be efficient if the molecule is coloured and hence able to absorb the visible light. This problem can be overcome by using unsaturated compounds, such as benzophenone (PhCOPh), that act as photosensitizers. Photosensitizers readily absorb light and can pass on the absorbed energy so as to cause dissociation of non-absorbing molecules.

$$\text{Energy (kJ mol}^{-1}) = \frac{N_A h c}{\lambda} = \frac{1.19 \times 10^5}{\lambda \text{ (nm)}}$$

N_A = Avogadro's number; h = Planck's constant; c = speed of light

Name	λ (nm)	Energy (kJ mol^{-1})
Visible light	780–380	153–313
Near or quartz light UV	380–200	313–595
Far or vacuum UV	200–10	595–11 900

Fig. 3.4

Photolysis has some important advantages over thermolysis. High-energy UV light can be used to cleave even very strong bonds that would otherwise require extremely high temperatures. In addition, as compounds can be irradiated with light of a particular energy, only certain bonds are broken and photolysis is a more selective, or 'cleaner', method of homolysis than thermolysis, which can lead to a number of side reactions. Photolysis can, however, require specialized apparatus (e.g. light sources, reactors) and, for UV experiments, quartz reaction vessels are usually necessary as Pyrex glass absorbs light strongly below 320 nm. (This is why a person can get heatstroke, but not sunburn, when sitting in a car with the windows shut.)

Not all of the energy absorbed by an organic molecule is used for photodissociation or breaking a bond. There are a number of pathways (including phosphorescence and fluorescence) by which an excited molecule can react, or alternatively use or lose its excess energy. The fraction of molecules that undergo homolysis as a result of excitation can be described by the primary quantum yield ϕ, where ϕ is the number of molecules of reactant consumed per photon of light absorbed; a photon is a definite amount (or quantum) of light energy (Fig. 3.5). This describes the efficiency of bond homolysis and the value of ϕ can be anywhere between zero and unity. When $\phi = 1$, every absorption causes homolysis of the bond. Alternatively, if the initial radical ($R^{1\bullet}$) leads to the formation of other radicals ($R^{2\bullet}$, $R^{3\bullet}$, etc.), as in a chain reaction (see Section 2.5), we can quote the overall quantum yield Φ for a particular product. This is the number of molecules of a product formed per photon of light absorbed by the reactant, and for some chain reactions this number can be greater than 10^6.

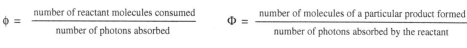

$$\phi = \frac{\text{number of reactant molecules consumed}}{\text{number of photons absorbed}} \qquad \Phi = \frac{\text{number of molecules of a particular product formed}}{\text{number of photons absorbed by the reactant}}$$

Fig. 3.5

3.3.1 Peroxides and diacylperoxides

Photolysis can be used to cleave the weak O—O bond of peroxides and diacylperoxides. The strength of the O—O bond in most of these molecules is between 125 and 155 kJ mol^{-1}, and this is equivalent to light of wavelength 770–950 nm. Visible light should therefore be more than capable of cleaving this bond. However, these molecules are stable in visible light because they are colourless (and therefore cannot absorb the light). Peroxides only start to absorb light in the UV region (up to approximately

350 nm); therefore, for efficient radical generation, these compounds need to be photolysed with UV light. Hydrogen peroxide, for example, readily dissociates (with a quantum yield of $\phi \approx 0.85$) to two hydroxyl radicals when irradiated with UV light ($\lambda = 254$ nm).

3.3.2 Azo compounds

The two weak C—N bonds of alkyl azo compounds (R—N=N—R) can be cleaved using UV light. Although the energy of visible light is sufficient for the cleavage of these bonds, like peroxides, they generally exhibit very weak absorption at wavelengths greater than 300 nm. AIBN is an exception, having a strong absorption band at 345 nm, but even this wavelength is in the near-UV region. Aromatic azo compounds have much stronger C—N bonds, which are difficult to cleave because the resulting aryl radicals are much less stable than alkyl radicals.

3.3.3 Halogens and halides

Irradiation of gaseous chlorine, bromine or iodine can lead to homolytic cleavage of the weak σ bonds (Fig. 3.6). The molecule is excited and, in the excited state, an electron occupies an antibonding σ^* orbital, which leads to weakening of the bond and dissociation. Visible light (of 500 nm) can be used to generate iodine atoms from molecular iodine, whilst the other halogens require light of shorter wavelength. Whereas temperatures above 200°C are required to generate chlorine atoms in the dark, irradiation of chlorine with UV light can lead to dissociation at room temperature. These photolyses are very efficient and dissociation of bromine (using UV light) has a primary quantum yield (ϕ) of unity.

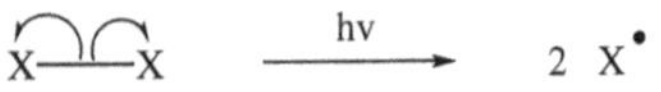

X	Bond dissociation energy (kJ mol^{-1})
Cl	245
Br	190
I	150

Fig. 3.6

Organohalogen derivatives, with weak C—X bonds (where X = halogen), can also undergo photolysis (Fig. 3.7). Alkyl chlorides absorb in the vacuum UV region (173 nm for CH_3Cl), while bromides or iodides absorb at longer wavelength (258 nm for CH_3I). Alkyl iodides with very weak C—I bonds (≈ 215 kJ mol^{-1}) are particularly susceptible to homolytic cleavage, but this is not always an efficient process, and the quantum yield (ϕ) for the homolysis of ethyl iodide is 0.31 in solution. The low value has been attributed to the formation of a diradical 'cage' species in which the radicals are surrounded by a shell of solvent molecules (see Section 4.6.1). Alkyl and iodine radicals that do not escape the solvent cage tend to recombine to regenerate the starting material. Vinyl and aryl iodides can also be photolysed using UV light, and this is a very useful and selective method for generating (particularly reactive) vinyl and aryl radicals. For polyhalogenated alkanes, the weakest bond is selectively cleaved and, for example, the weaker C—Br bond (234 kJ mol^{-1}) in $BrCCl_3$ undergoes homolysis to Br$^\bullet$ and $^\bullet CCl_3$ radicals. Chlorofluorocarbons (CFCs), such as CF_2Cl_2 (Freon 12), undergo a similar photolysis in the stratosphere, and the resulting chlorine radicals contribute to the destruction of the ozone layer.

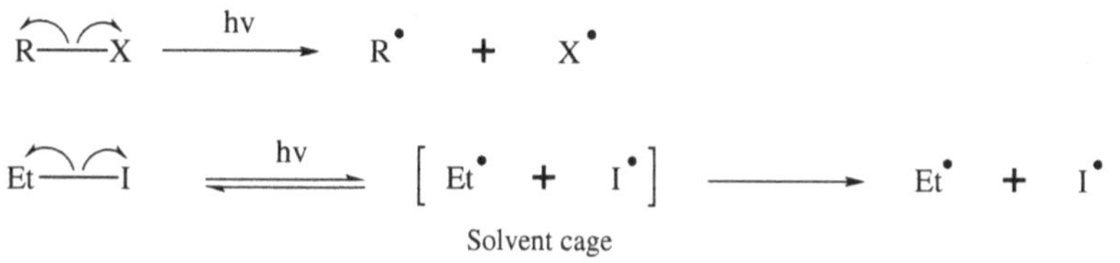

Fig. 3.7

Oxygen- and nitrogen-centred radicals can also be produced by photolysis of weak (≈ 190–210 kJ mol^{-1}) heteroatom–halogen bonds (Fig. 3.8). The most common precursors are hypochlorites (RO—Cl) and N-chloroamines (R_2N—Cl), which can decompose to alkoxyl (RO$^\bullet$) and aminyl ($R_2N^\bullet$) radicals, respectively. Acidic conditions are often required for efficient reaction of N-chloroamines, and this leads to the formation of radical cations, $R_2NH^{+\bullet}$, which are known as aminium radicals (see Chapter 11).

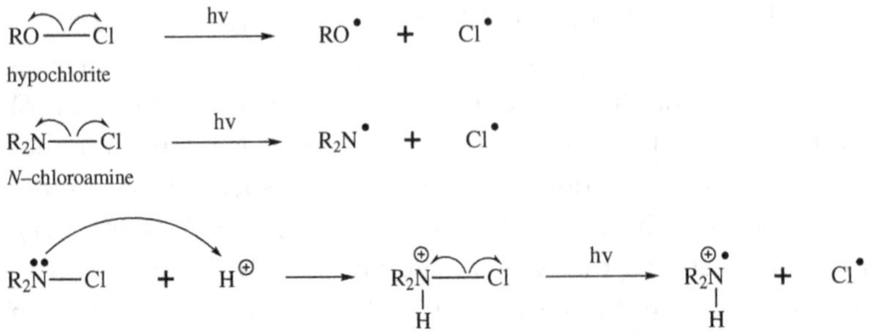

$$RO \overset{\frown}{\frown} Cl \xrightarrow{h\nu} RO^{\bullet} + Cl^{\bullet}$$

hypochlorite

$$R_2N \overset{\frown}{\frown} Cl \xrightarrow{h\nu} R_2N^{\bullet} + Cl^{\bullet}$$

N–chloroamine

$$R_2\ddot{N}-Cl + H^{\oplus} \longrightarrow \overset{\oplus}{R_2N} \overset{\frown}{\frown} Cl \xrightarrow{h\nu} \overset{\oplus}{R_2N}{}^{\bullet} + Cl^{\bullet}$$
$$\underset{H}{|} \qquad\qquad \underset{H}{|}$$

Fig. 3.8

3.3.4 Nitrites

Organic nitrites (R—O—N=O) absorb light in the UV region (around 360 nm), leading to an excited state resulting from an n→π* transition. The excited state reacts by cleaving the weak (≈ 175 kJ mol^{-1}) RO—NO single bond to produce an alkoxyl radical (RO$^{\bullet}$) together with nitric oxide ($^{\bullet}$NO) (Fig. 3.9). Nitric oxide is an important radical in human biology, and white blood cells use nitric oxide to kill foreign invaders (e.g. bacteria, viruses and parasites).

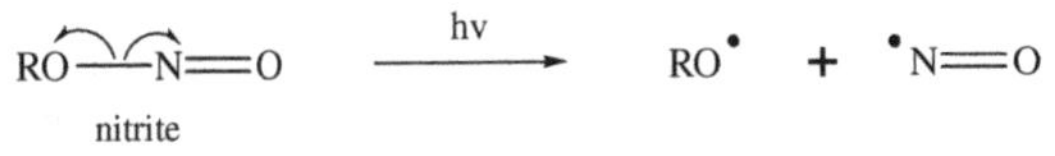

$$RO \overset{\frown}{\frown} N{=}O \xrightarrow{h\nu} RO^{\bullet} + {}^{\bullet}N{=}O$$

nitrite

Fig. 3.9

3.3.5 Organometallics

Metal-centred radicals can be prepared by the photolysis of weak metal–metal or metal–carbon bonds. The photolysis of alkylditins (R_3Sn—SnR_3) is particularly important because the resultant tin radicals have found numerous applications in organic synthesis (see Section 5.3.1). The Sn—Sn bond in these molecules is weaker than the Sn—C bond (by around 50 kJ mol^{-1}) and can be homolytically cleaved on irradiation with UV light (or less commonly by thermolysis) (Fig. 3.10). Absorption of light leads to a σ→σ* transition and the excited state has a weaker Sn—Sn bond. Other metal radicals can also be produced on photolysis of dimeric species, and the strength of the metal–metal bond is influenced by the nature of the ligands; large bulky ligands can weaken the σ bond and so facilitate radical formation.

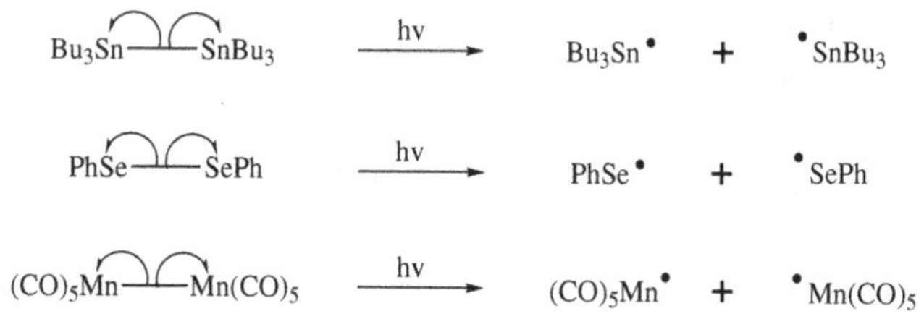

$$Bu_3Sn\text{---}SnBu_3 \xrightarrow{h\nu} Bu_3Sn^\bullet + {}^\bullet SnBu_3$$

$$PhSe\text{---}SePh \xrightarrow{h\nu} PhSe^\bullet + {}^\bullet SePh$$

$$(CO)_5Mn\text{---}Mn(CO)_5 \xrightarrow{h\nu} (CO)_5Mn^\bullet + {}^\bullet Mn(CO)_5$$

Fig. 3.10

An alternative method for generating metal radicals involves the photolysis of weak metal–carbon bonds. The Pb—C bond of alkyllead compounds can undergo homolysis (see Sections 1.2 and 1.4), while organocobalt and organomercury compounds are particularly prone to homolytic decomposition. The cobalt–carbon covalent bond in alkylcobalt(III) species is very weak (85–125 kJ mol^{-1}) and, when irradiated with light of wavelength between 500 and 300 nm, it is cleaved to give a carbon-centred radical and a cobalt(II) radical (Fig. 3.11). The ligands on cobalt and the organic substituent (R) both affect the strength of the bond; the more stable the product radical (R$^\bullet$), the weaker the bond. This method of radical generation is exploited by vitamin B$_{12}$ to catalyse a number of natural rearrangement reactions—these are triggered by homolysis of a cobalt(III)–carbon bond in the vitamin.

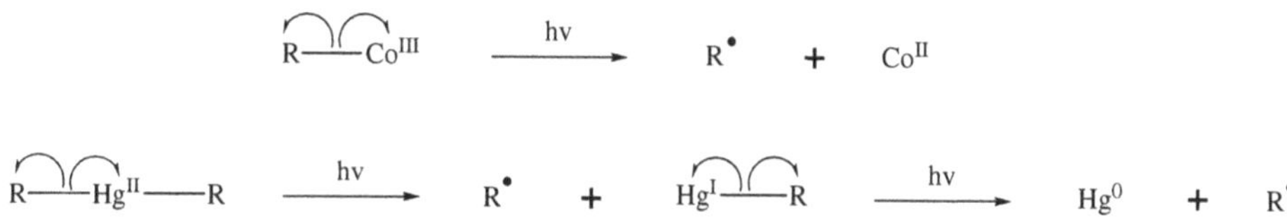

$$R\text{---}Co^{III} \xrightarrow{h\nu} R^\bullet + Co^{II}$$

$$R\text{---}Hg^{II}\text{---}R \xrightarrow{h\nu} R^\bullet + Hg^{I}\text{---}R \xrightarrow{h\nu} Hg^0 + R^\bullet$$

Fig. 3.11

Dialkylmercury(II) compounds, R$_2$Hg, contain two very weak Hg—C bonds and readily decompose to mercury(0) and two alkyl radicals (Fig. 3.11). The bonds are so weak (typically 60 kJ mol^{-1}) that, in some cases, compounds need to be stored in the dark at low temperature to prevent premature photolysis or thermolysis. An exception is dimethylmercury(II), which has a much stronger Hg—Me bond (121 kJ mol^{-1}) and, like diarylmercury(II) compounds, slowly decomposes because the methyl (or aryl) radical is so unstable. Decomposition of the intermediate alkylmercury(I) species is very fast and these intermediates are most often generated from organomercury(II) hydrides (RHg^{II}H). These compounds contain a very weak Hg—H bond and are so unstable that they usually cannot be isolated and so are generated *in situ* (see Section 5.3.2).

Metal-centred radicals produced on homolysis of metal–metal, metal–carbon and metal–hydrogen bonds have been identified as intermediates in a number of important metal-promoted organic reactions. These include metal-catalysed hydrogenations and some reactions of Grignard (R—MgX) reagents, where alkyl radicals are derived from homolysis of the magnesium–carbon bond.

3.3.6 Carbonyls

Aldehydes and ketones have a weak UV absorption (around 270–300 nm) which causes one electron from an oxygen lone pair in a non-bonding orbital (n) to be promoted to a higher energy antibonding π orbital (π*). This is described as an n→π* transition. In this (singlet) excited state, the molecule can lose energy to form an excited (triplet) state in which the electrons of the C=O π bond are no longer spin paired, i.e. the two unpaired electrons have the *same* spin. This singlet to triplet process is known as intersystem crossing. The triplet state is usually represented as shown in Fig. 3.12, and it is this diradical that can participate in radical reactions.

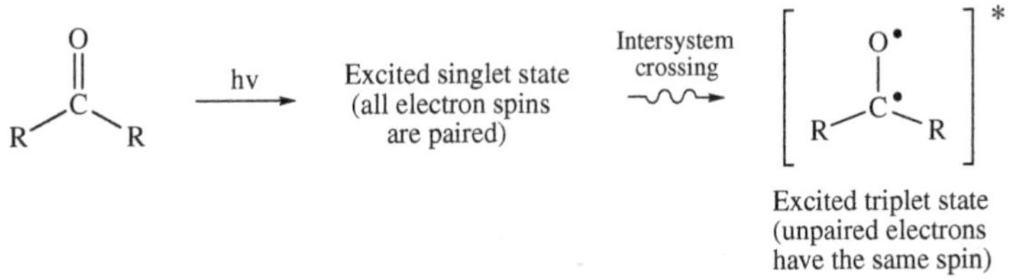

Fig. 3.12

With regard to other photolytic methods, the number of diradicals produced depends on how efficiently the UV light is absorbed. This is measured by the molar extinction coefficient or molar absorptivity (ε), which is characteristic of the molecule. The more strongly a molecule absorbs light of a particular wavelength, the higher the value of ε. For most saturated carbonyls the absorption between 270 and 300 nm has $\varepsilon \approx 20$, but if the C=O group is conjugated a slightly stronger absorption (at longer wavelength) can be observed. For acetophenone (PhCOCH$_3$), for example, the n→π* transition occurs at $\lambda_{max} \approx 320$ nm with $\varepsilon \approx 60$ (in ethanol). The efficiency of intersystem crossing is also important for diradical formation and this depends on the structure of the molecule. For example, singlet to triplet conversion is particularly efficient for aromatic carbonyls, ArC(O)R, as the carbon-centred radical can be delocalized onto the adjacent benzene ring. The triplet diradical should therefore be more stable and longer lived, which explains why photochemical reactions of aryl ketones are generally the result of triplet state processes.

3.4 Radiolysis and sonolysis

Other sources of energy can be used to cleave bonds and generate radicals. When organic molecules are exposed to high-energy radiation, such as X-rays or γ-radiation (from, for example, ^{60}Co or ^{137}Cs sources), radicals are formed, but their subsequent reactions can be complex and unselective. The mechanism involves the ejection of an electron from the precursor to form a radical cation that decomposes to a radical and a cation (Fig. 3.13). Reactions in water generate reactive hydroxyl radicals which can then react with naturally occurring molecules to mimic the effects of radiation on biological systems (see Section 1.3). Radiolysis is an important method of radical generation in the food industry. Irradiation of food (with X-rays or γ-rays) generates reactive radicals that kill contaminating microorganisms, and has been shown to be an effective and versatile method of preserving food.

Covalent bonds can also be broken by high-frequency sound waves (ultrasound) using a technique known as sonolysis. As the sound waves pass through a solution, bubbles are formed in a process known as cavitation and these collapse to release large amounts of energy. The high temperatures and pressures that are generated locally can lead to the homolysis of molecules in solution. For water, this process produces hydrogen and hydroxyl radicals, whilst in chloroform the carbon–chlorine bond is preferentially cleaved (Fig. 3.13).

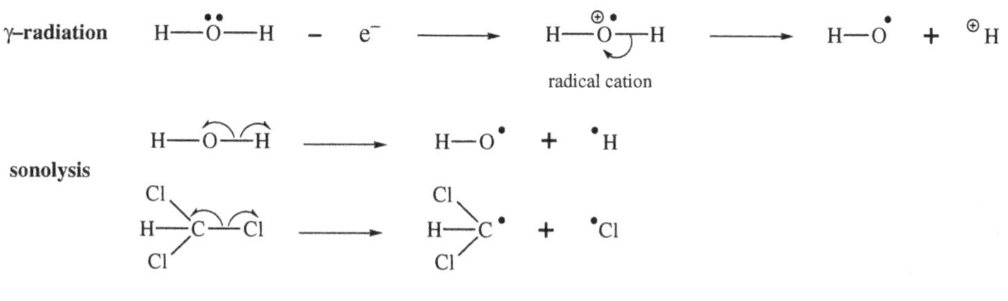

Fig. 3.13

3.5 Electron transfer

Radicals are formed when an electron is transferred to, or from, a molecule with (only) paired electrons. A neutral molecule can be converted to a radical by addition or loss of one electron (Fig. 3.14). If we add an electron, we generate a radical anion and this usually fragments to a radical and an anion (see Chapter 10). The 'extra' electron in the radical anion occupies a high-energy antibonding orbital and this weakens the structure, commonly leading to the cleavage of a bond in a saturated molecule. If we lose an electron, a radical cation is produced which fragments to a radical and a cation (Fig. 3.14; see Chapter 11). The bond is weakened in the radical cation because the electron is removed from a (low-energy) bonding orbital. Alternatively, radicals can be generated from (non-radical) anions by loss of an electron, or from (non-radical) cations by addition of an electron. These redox reactions require (one-electron) oxidizing and reducing agents, and metal ions which change their oxidation state by one are most commonly employed. This is a mild method and reactions can be conducted at room temperature or below and, unlike thermolysis or photolysis, this leads to the generation of only one organic radical (rather than a pair). The method is also selective and the intermediate radical anion or cation will fragment so as to give the most stable radical and ionic products.

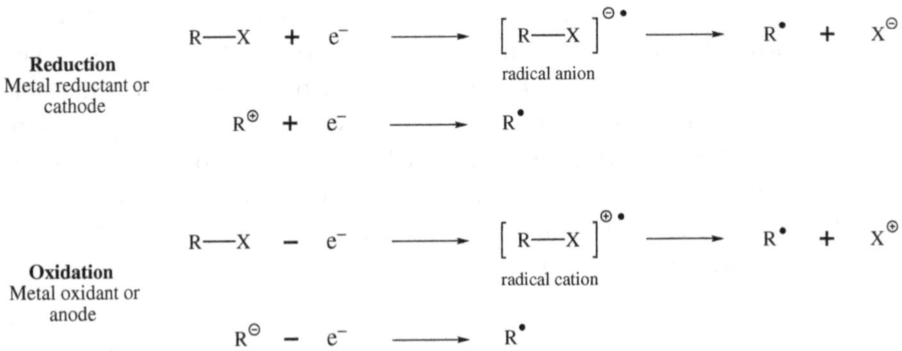

Reduction
Metal reductant or cathode

$$R\text{---}X \;+\; e^- \;\longrightarrow\; \left[\, R\text{---}X \,\right]^{\ominus \bullet} \;\longrightarrow\; R^{\bullet} \;+\; X^{\ominus}$$

radical anion

$$R^{\oplus} \;+\; e^- \;\longrightarrow\; R^{\bullet}$$

Oxidation
Metal oxidant or anode

$$R\text{---}X \;-\; e^- \;\longrightarrow\; \left[\, R\text{---}X \,\right]^{\oplus \bullet} \;\longrightarrow\; R^{\bullet} \;+\; X^{\oplus}$$

radical cation

$$R^{\ominus} \;-\; e^- \;\longrightarrow\; R^{\bullet}$$

Fig. 3.14

Single-electron transfer can also be carried out in an electrolytic cell. The electron transfer occurs between a species in solution and an electrode (electron conductor) made from a metal or carbon. Electrolysis of neutral molecules or ions can produce radicals. Neutral molecules will be reduced to radical anions at the (negatively charged) cathode and oxidized to radical cations at the (positively charged) anode. Alternatively, anions will be electrostatically attracted to the anode and lose an electron, while cations will gain an electron at the cathode.

3.5.1 Reduction

For radical generation by reduction, we need a precursor able to accept an electron and a reagent willing to donate an electron. We can determine the likelihood of these processes occurring from the oxidation and reduction potentials; these reflect the ease with which a species is oxidized or reduced. A low, preferably negative, standard reduction potential[1] ($E^{\ominus}$) indicates a good reducing agent. The most frequently used reducing reagents are metal ions that readily lose an electron and so increase their oxidation state by one (Fig. 3.15). Group I (or alkali) metals, including lithium, sodium and potassium, are among the most powerful reducing agents, as loss of an electron produces a stable cation which is isoelectronic with the noble gases.

$$M^+ \ + \ e^- \longrightarrow M \qquad M\ (E^{\ominus}/V) : \text{Li } (-3.045), \text{Na } (-2.711), \text{K } (-2.924)$$

(By convention, the half-cell is quoted in terms of a reduction and negative values indicate the reverse process is favoured)

Fig. 3.15

We do not often know the reduction potentials of organic compounds because values for only a limited number of relatively simple molecules have been determined. However, we might expect a molecule to be readily reduced if, after accepting an electron, it decomposes to form a stable radical and/or anion.

3.5.1.1 Halides

The addition of an electron to an alkyl or aryl halide (R—X) can lead to radical generation (Fig. 3.16). Reductions traditionally use alkali metals to produce an intermediate radical anion that collapses to an alkyl or aryl radical (R•). The electronegative halogen substituent (X) strongly attracts the electrons and ensures that a halide anion rather than halogen atom is produced on bond cleavage. A number of milder metal-mediated methods have since been developed. Samarium(II) iodide is a good electron donor (see Section 5.4.1) and readily reduces alkyl iodides, while copper(I), iron(II) and ruthenium(II) complexes have all been used to generate radicals from substituted alkyl chlorides.

---

[1] Using the standard hydrogen electrode as a reference electrode.

$$M^n \ + \ R\!-\!X \ \longrightarrow \ M^{n+1} \ + \ \left[R\!-\!X \right]^{\ominus \bullet} \ \longrightarrow \ R^{\bullet} \ + \ X^{\ominus}$$

alkyl halide (X = Cl, Br or I)

M ($E^{\ominus}$/V)	
Sm^{3+}/Sm^{2+}	(−1.00)
Cu^{2+}/Cu^{+}	(0.15)
Ru^{3+}/Ru^{2+}	(0.25)
Fe^{3+}/Fe^{2+}	(0.77)

Fig. 3.16

3.5.1.2 Peroxides

Metal reducing agents, including copper(II), iron(II), chromium(II) and cobalt(II), can react with peroxides to generate oxygen-centred radicals. The metal rapidly donates an electron to the antibonding (σ^*) orbital of the oxygen–oxygen bond and the radical anion will fragment to an alkoxyl radical ($RO^{\bullet}$) and alkoxide anion (RO^-) (Fig. 3.17). Hydroperoxides decompose to give the more stable alkoxyl (rather than hydroxyl) radical, together with hydroxide. To make hydroxyl radicals, we need to use hydrogen peroxide (H_2O_2) and, when iron(II) is the reductant, this is known as Fenton's reaction (Fig. 3.17).

$$M^n \ + \ RO\!-\!OR \ \longrightarrow \ M^{n+1} \ + \ \left[RO\!-\!OR \right]^{\ominus \bullet} \ \longrightarrow \ RO^{\bullet} \ + \ RO^{\ominus}$$

$$Fe^{2+} \ + \ HO\!-\!OH \ \longrightarrow \ Fe^{3+} \ + \ HO^{\bullet} \ + \ HO^{\ominus} \qquad \text{Fenton's reaction}$$

Fig. 3.17

3.5.1.3 Diazonium salts

Radicals can also be generated from arenediazonium salts, ArN_2^+ (Fig. 3.18). These cations are reduced by, for example, copper(I) salts or organic reductants (which readily lose an electron to form a stable radical cation) such as tetrathiofulvalene (TTF), and the addition of one electron produces an intermediate nitrogen-centred radical. This rapidly fragments to give nitrogen gas, which is an excellent leaving group (cf. azo compounds, Section 3.2.2), together with an aryl radical. The aryl radicals can react further so as to introduce substituents onto the benzene ring. This has been exploited in a number of 'named' reactions, including the Sandmeyer reaction (for preparing aryl bromides and chlorides), the Meerwein arylation (for alkenylation) and the Pschorr reaction (for intramolecular arylation).

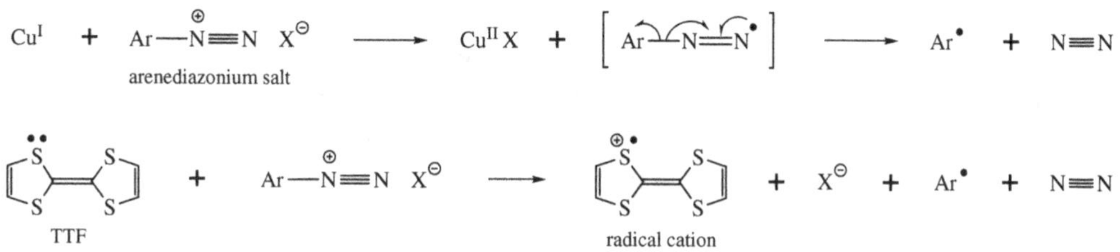

Fig. 3.18

3.5.2 Oxidation

For radical generation by oxidation, we need a precursor able to donate an electron and a reagent willing to accept an electron. A high standard reduction potential ($E^{\ominus}$) indicates a good oxidizing agent. The most frequently used oxidizing reagents are metal ions that readily accept an electron and so decrease their oxidation state by one (Fig. 3.19). Strongly oxidizing metal ions include silver(II), lead(IV), cerium(IV) and manganese(III).

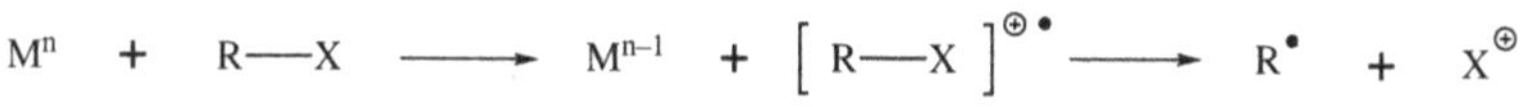

M ($E^{\ominus}$/V)	
Ag^{2+}/Ag^{+}	(1.98)
Pb^{4+}/Pb^{3+}	(1.69)
Ce^{4+}/Ce^{3+}	(1.61)
Mn^{3+}/Mn^{2+}	(1.54)

Fig. 3.19

We do not often know the oxidation potentials of organic compounds because values for only a limited number of relatively simple molecules have been determined. However (in accord with Hammond's postulate), we might expect a molecule to be readily oxidized if, after donating an electron, it decomposes to form a stable radical and/or cation.

3.5.2.1 Carboxylic acids and alcohols

Oxidation of carboxylic acids with a metal salt generates radical cations (Fig. 3.20). The radical cation fragments to give a carboxyl (or acyloxyl) radical, stabilized by resonance delocalization, together with a proton. A number of strongly oxidizing metal ions can be used, including silver(II). Silver(II) is frequently generated *in situ* by oxidation of silver(I) using sodium persulfate or peroxydisulfate ($Na_2S_2O_8$). This method has the advantage that only a catalytic amount of the expensive silver(II) salt is required—reaction with the acid produces silver(I) which is reoxidized by the persulfate. Persulfates are cheap, widely used oxidizing agents which readily accept a total of two electrons to form two sulfate ions (SO_4^{2-}) by cleavage of the weak oxygen–oxygen bond (cf. peroxides).

$$M^n \; + \; RCO_2H \longrightarrow M^{n-1} \; + \; \left[RCO_2H \right]^{\oplus \bullet} \longrightarrow RCO_2^{\bullet} \; + \; H^{\oplus}$$

carboxyl or
acyloxyl radical

$$Ag^+ \; + \; {}^{\ominus}O_3SO-OSO_3{}^{\ominus} \longrightarrow Ag^{2+} \; + \; SO_4^{\ominus \bullet} \; + \; SO_4^{2\ominus}$$

persulfate

$$Ag^+ \; + \; SO_4^{\ominus \bullet} \longrightarrow Ag^{2+} \; + \; SO_4^{2\ominus}$$

Fig. 3.20

A similar reaction occurs when alcohols are treated with oxidizing metals. An electron is abstracted from an oxygen lone pair and the resulting radical cation fragments to give an alkoxyl radical and a proton (Fig. 3.21).

$$M^n \; + \; ROH \longrightarrow M^{n-1} \; + \; \left[ROH \right]^{\oplus \bullet} \longrightarrow RO^{\bullet} \; + \; H^{\oplus}$$

Fig. 3.21

3.5.2.2 Carboxylates and phenoxides

Carboxylate salts are oxidized at the anode of an electrochemical cell in the Kolbe reaction. The carboxylate loses an electron at the positive electrode (anode) to generate an acyloxyl radical (Fig. 3.22). This process can also be used to generate phenoxyl radicals from phenoxides.

$$RCO_2^{\ominus} \; - \; e^- \longrightarrow RCO_2^{\bullet}$$

$$ArO^{\ominus} \; - \; e^- \longrightarrow ArO^{\bullet}$$

Fig. 3.22

3.5.2.3 Alkylaromatics

Reaction of alkylbenzene derivatives with strong oxidizing agents, such as cerium(IV), can promote radical generation. For toluene (methylbenzene), an electron is removed from the π system of the benzene ring and the radical cation fragments to give a benzyl radical and a proton (Fig. 3.23). Oxidation of the carbon atom adjacent to the benzene ring is favoured because a stable, resonance-delocalized, benzyl radical is formed.

$$Ce^{4+} \quad + \quad Ph-CH_3 \quad \longrightarrow \quad Ce^{3+} \quad + \quad \left[Ph-CH_3\right]^{\oplus \bullet} \quad \longrightarrow \quad PhCH_2^{\bullet} \quad + \quad H^{\oplus}$$

Fig. 3.23

3.5.2.4 Carbonyls

Carbonyl compounds can undergo a one-electron oxidation with manganese(III) salts (Fig. 3.24). The exact mechanism of the reaction is still under investigation, but the rate of radical generation is known to be fastest for carbonyl compounds that will readily enolize. This includes 1,3-dicarbonyl compounds, for which the enol tautomer is stabilized by conjugation and by intramolecular hydrogen bonding. When substrates of this type are reacted with manganese(III), a resonance-stabilized radical is generated at the α-carbon, together with manganese(II). This may involve an intermediate manganese(III) enolate which could be formed from reaction of the enol with, for instance, manganese(III) acetate (ethanoate). The resulting enolate could then decompose by homolysis of the O—Mn bond to give manganese(II) and the resonance-stabilized radical (see Section 5.4.3).

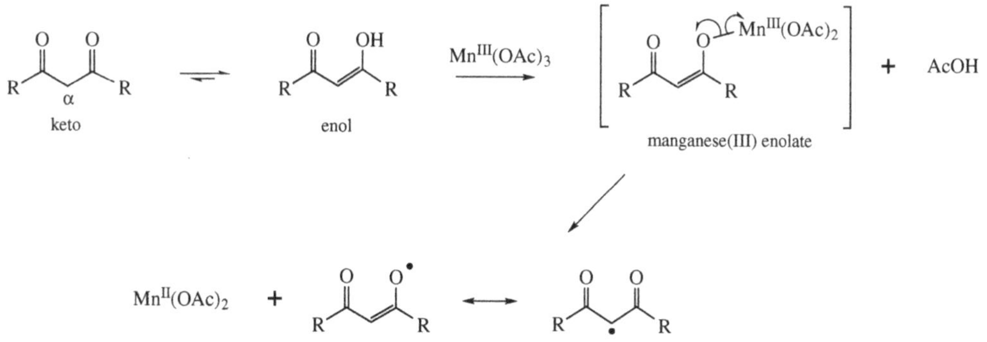

Fig. 3.24

3.6 Summary

There are a variety of methods available for making radicals. For radical generation, we can see that precursors with weak bonds are required as these can be easily and selectively cleaved. Heating a compound is probably the simplest way to make radicals, and molecules with very weak bonds can be cleaved at temperatures below 150°C. Other methods, such as photolysis and electron transfer, allow radicals to be prepared under milder reaction conditions (e.g. room temperature) and can lead to more selective radical generation.

Further reading

Dalko, P.I. (1995) Redox induced radical and radical ionic carbon–carbon bond forming reactions. *Tetrahedron*, **51**, 7579–7653.

Iqbal, J., Bhatia, B. & Nayyar, N.K. (1994) Transition metal-promoted free-radical reactions in organic synthesis: the formation of carbon–carbon bonds. *Chemical Reviews*, **94**, 519–564.

Marcus, R.A. (1993) Electron transfer reactions in chemistry: theory and experiment (Nobel lecture). *Angewandte Chemie, International Edition in English*, **32**, 1111–1222.

Russell, G.A. (1987) Free radical chain processes in aliphatic systems involving an electron transfer reaction. *Advances in Physical Organic Chemistry*, **23**, 271–322.

CHAPTER 4

Radical Reactions

4.1 Introduction

When radicals react, they can form more stable non-radical products in which the electrons are paired. The simplest reaction involves the combination (or coupling) of two radicals to form a new two-electron covalent bond (Fig. 4.1). Alternatively, radicals can lose an electron to form cations (oxidation) or accept an electron to form anions (reduction). These ionic intermediates can then react, for example, with nucleophiles (Nu⁻) or electrophiles (E⁺), respectively, to form neutral, non-radical molecules. As all of these reactions destroy the radical intermediates, they are known as termination reactions.

Combination	$R^\bullet$	$+$	$^\bullet R$	$\longrightarrow$	$R\!-\!R$	
Oxidation	$R^\bullet$	$-$	e^-	$\longrightarrow$	$R^\oplus$ $\xrightarrow{Nu^\ominus}$	$R\!-\!Nu$
Reduction	$R^\bullet$	$+$	e^-	$\longrightarrow$	$R^\ominus$ $\xrightarrow{E^\oplus}$	$R\!-\!E$

Termination Reactions

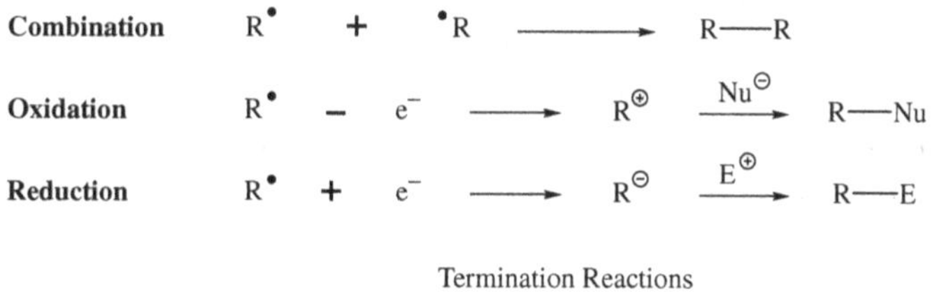

Fig. 4.1

Radicals can also react to produce new and more stable radicals before termination reactions can take place. These reactions, which have a radical starting material ($R^{1\bullet}$) and a radical product ($R^{2\bullet}$), are known as propagation reactions. There are a number of these reactions which can be divided into four main groups: (i) radical abstraction of an atom from a non-radical molecule; (ii) radical addition to a non-radical molecule; (iii) radical fragmentation to give a new radical by loss of a non-radical molecule; and (iv) radical rearrangements (Fig. 4.2). In each case, one radical is produced ($R^{2\bullet}$) as another is destroyed ($R^{1\bullet}$).

Abstraction	$R^{1\bullet}$ + X—R^2 $\longrightarrow$ R^1—X + $R^{2\bullet}$
Addition	$R^{1\bullet}$ + Non-radical $\longrightarrow$ $R^{2\bullet}$
Fragmentation	$R^{1\bullet}$ $\longrightarrow$ Non-radical + $R^{2\bullet}$
Rearrangement	$R^{1\bullet}$ $\longrightarrow$ $R^{2\bullet}$

Propagation Reactions

Fig. 4.2

In Chapter 2, we saw that not all radicals have the same stability. Their stability is linked to reactivity, and radicals stabilized by conjugation, for example, tend to be less reactive because the electron density is spread over the molecule and not concentrated on one atom. Reactivity is also linked to steric effects because bulky substituents can shield the radical making it less reactive. The electronic (mesomeric and inductive) effects of the substituents are also important as the 'polar character' of radicals influences their reactivity. As a consequence, not all radicals will react at the same rate or even by the same reaction pathway. To predict which pathway the radical will choose, we need to know the relative rates of the competitive (propagation and termination) processes. In this chapter, the most common propagation and termination reactions are introduced, and this is followed by a discussion of the rates of these processes.

4.2 Propagation reactions

4.2.1 Radical abstraction reactions

Radical abstraction or transfer reactions generally involve the removal of an atom (such as hydrogen or a halogen) from a non-radical precursor. These reactions can be bimolecular (intermolecular) or unimolecular (intramolecular) processes. The mechanism of these reactions involves the attack of a radical ($R^{1\bullet}$) at the σ bond of the atom undergoing abstraction to form a new, more stable, product radical ($R^{2\bullet}$) (Fig. 4.3). For efficient radical generation, we need precursor molecules with a weak σ bond; the weaker the σ bond in the precursor, the more stable the product radical ($R^{2\bullet}$). These reactions lead to the formation of a stronger σ bond (R^1—X) in the product than that broken in the precursor (R^2—X).

$R^{1\bullet}$ + X—R^2 $\longrightarrow$ R^1—X + $R^{2\bullet}$ X = H or halogen

Fig. 4.3

4.2.2 Intermolecular (S_H2) reactions

The intermolecular reaction is also referred to as a substitution or displacement reaction and given the symbol S_H2—substitution, homolytic, bimolecular (Fig. 4.4). This substitution process is believed to involve the attack of the radical at an angle of approximately 180° to the R^2—X bond as this maximizes the interaction of the radical orbital and the vacant σ^* orbital of the bond which is broken (see Section 4.7.4). The atom to be abstracted (X) is positioned between the R^1 and R^2 groups in a linear transition state (designated ‡), and the R^1—X bond is forming as the R^2—X bond is breaking (in a *concerted* process).

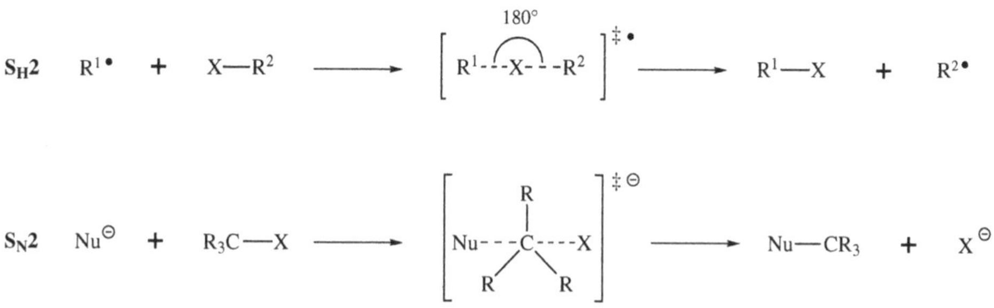

Fig. 4.4

The S_H2 mechanism is related to the ionic S_N2 (substitution, nucleophilic, bimolecular) reaction, which proceeds via a transition state in which the nucleophile and leaving group (X) are both partially bonded to the central carbon. It is very rare, however, for radical abstractions to take place by the radical attacking a carbon atom and the most common atoms to be attacked (and therefore abstracted) are hydrogen and halogen atoms.

Hydrogen atom abstraction can occur, for example, when heteroatom radicals, such as halogen or oxygen-centred radicals, are reacted with alkanes. Alkoxyl radicals (RO•), for example, abstract a hydrogen atom so as to form an alcohol with a strong O—H bond (Fig. 4.5). Hydrogen atoms are more readily abstracted from tertiary rather than secondary or primary C—H bonds, and this is consistent with the bond strengths and the greater stability of tertiary radicals (see Section 2.4.1.1). However, the selectivity of the hydrogen atom abstraction also depends on the nature of the radical; bromine radicals (Br•), for example, react more selectively with alkanes than do chlorine radicals (Cl•) (see Section 4.7.1).

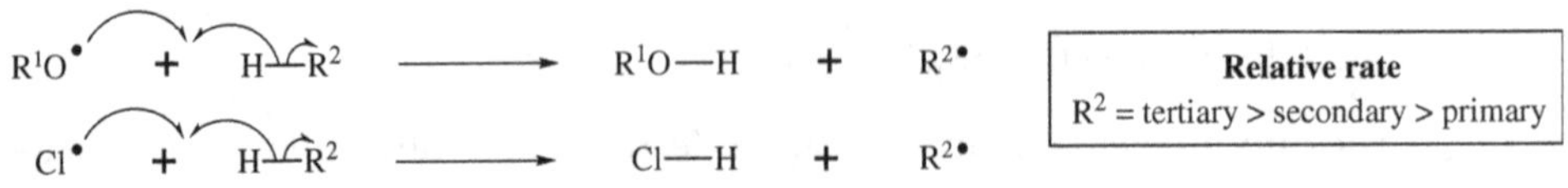

$$R^1O^\bullet \quad + \quad H\frown R^2 \quad \longrightarrow \quad R^1O{-}H \quad + \quad R^{2\bullet}$$

$$Cl^\bullet \quad + \quad H\frown R^2 \quad \longrightarrow \quad Cl{-}H \quad + \quad R^{2\bullet}$$

| **Relative rate** |
| R^2 = tertiary > secondary > primary |

Fig. 4.5

Tin, silicon or germanium radicals are often used to abstract halogen atoms (Fig. 4.6). The driving force for these reactions is the formation of a strong metal–halogen bond at the expense of a weaker carbon–halogen bond (see Sections 5.3.1 and 5.3.3). Tertiary halides react more quickly than primary halides, and the weak carbon–iodine bond ensures that alkyl iodides react more quickly than alkyl bromides, chlorides or fluorides.

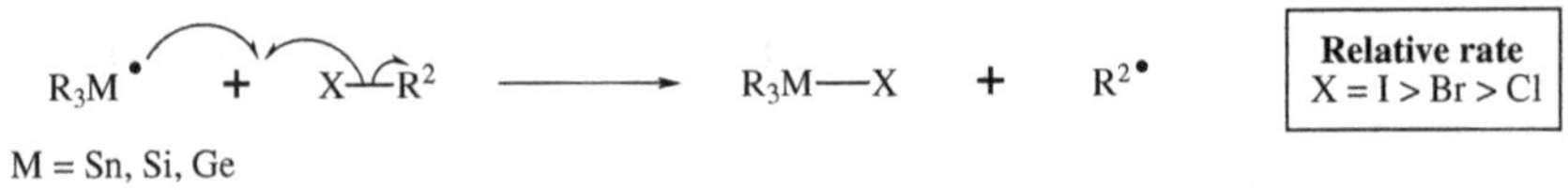

$$R_3M^\bullet \quad + \quad X\frown R^2 \quad \longrightarrow \quad R_3M{-}X \quad + \quad R^{2\bullet}$$

| **Relative rate** |
| $X = I > Br > Cl$ |

M = Sn, Si, Ge

Fig. 4.6

4.2.3 Intramolecular (S_Hi) reactions

Atom abstractions can also occur intramolecularly and, as for inter-molecular atom transfer, a linear transition state is preferred for S_Hi (substitution, homolytic, intramolecular) reactions. This is restrictive for intramolecular reactions, and atom transfers usually only occur via six- or seven-membered transition states as these can adopt an approximately linear geometry (Fig. 4.7). These are called 1,5- and 1,6-atom transfers as the atom moves from position 1 to position 5 or 6 on the chain. For atom transfers below 1,5-, the transition state contains five or fewer atoms and the ring is too small to adopt an approximately linear geometry. Atom transfers above 1,6- are less common because, as the chain becomes longer, the radical and atom (to be transferred) move further apart and so they are less likely to react.

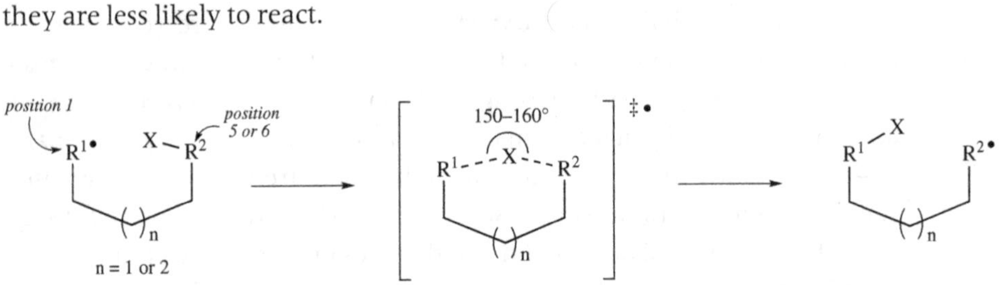

Fig. 4.7

The most common intramolecular reactions involve hydrogen atom transfer. Reactive oxygen-, nitrogen- or carbon-centred radicals can react with a C—H bond, leading to the formation of a more stable carbon-centred radical and a stronger σ bond (e.g. a strong O—H or N—H bond). These unimolecular reactions are normally faster than bimolecular atom transfers because, when a molecule reacts with 'itself', there is no change in (translational) entropy. For intermolecular reactions, however, the two reacting molecules must be briefly joined, resulting in a (temporary) decrease in entropy, and this is unfavourable.

4.3 Addition reactions

Carbon-centred radicals will add to a variety of unsaturated compounds to form new carbon–carbon bonds. These reactions proceed to form a new strong σ bond (C—C) at the expense of a weaker π bond (e.g. the π bond in an alkene). Reaction with alkynes and particularly alkenes is common, and this type of reaction is used in industry to make polymers (e.g. polystyrene from styrene, $PhCH{=}CH_2$). Examples of both inter- and intramolecular reactions are known and the rates of addition have been shown to depend on the nature of the radicals and the π systems.

4.3.1 Intermolecular addition to alkenes and alkynes

The addition of carbon radicals to alkenes is usually energetically favoured as a new C—C σ bond (≈ 370 kJ mol^{-1}) is formed at the expense of a weaker carbon–carbon π bond (≈ 235 kJ mol^{-1}). The radical is believed to approach the double bond at an angle of approximately 107°, and the p orbital interacts with either the π or π* orbital (see Sections 2.4.3 and 4.7.5) to form a new sp^3 carbon (with bond angles of $\approx 109°$) with an unpaired electron on an adjacent carbon atom. With unsymmetrical alkenes, the radical predominantly attacks at the least hindered position of the double bond (Fig. 4.8). When a radical adds to the 'end' of a monosubstituted alkene, a secondary radical is produced, whilst attack at the more hindered site produces a primary radical. The regioselectivity was initially explained by the greater stability of the secondary (over the primary) radical, but it now appears that steric factors are more important; the radical attacks the less substituted end of the double bond simply because this is less crowded and more accessible.

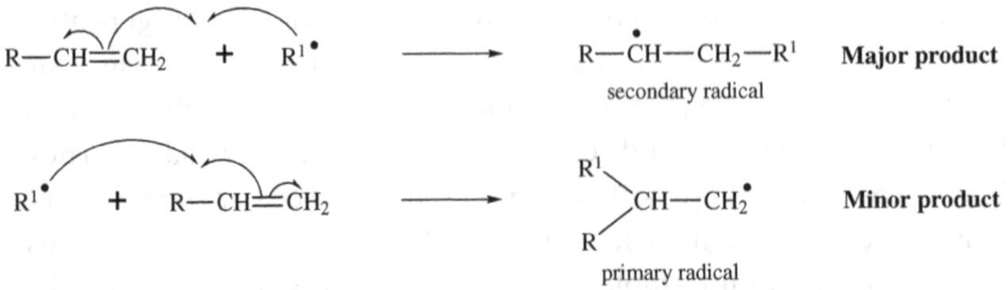

Fig. 4.8

This regioselectivity explains the anti-Markownikoff (anti-Markovnikov) addition of hydrogen bromide to alkenes in the presence of peroxide (Fig. 4.9). In the absence of peroxide, H—Br adds to a monosubstituted alkene in a heterolytic (non-radical) reaction to form predominantly the secondary halide (in a Markownikoff addition). This is because the proton adds to the carbon bearing more hydrogens to form the more stable (secondary rather than primary) carbocation intermediate. However, when peroxide (RO—OR) is added, alkoxyl radicals (RO•) are formed which abstract a hydrogen atom from H—Br (to form a strong O—H bond). The resulting bromine radical (Br•) adds selectively to the least hindered end of the alkene to give a secondary radical. This secondary radical abstracts a hydrogen atom from H—Br (to form a strong carbon–hydrogen bond) to give the primary halide (or anti-Markownikoff) product in a chain reaction.

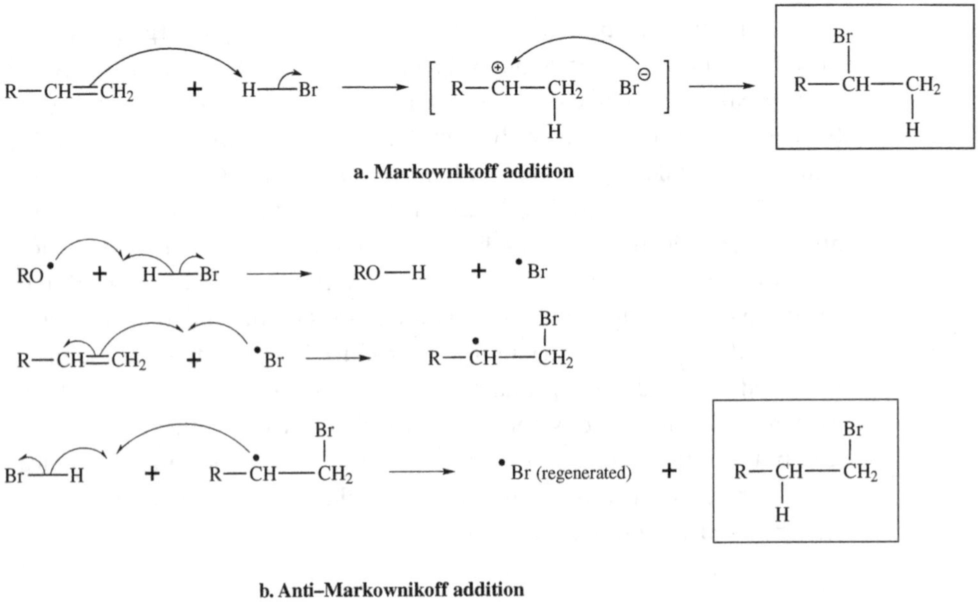

a. Markownikoff addition

b. Anti–Markownikoff addition

Fig. 4.9

Carbon radicals also add to the less hindered end of alkynes to form new carbon–carbon bonds (Fig. 4.10). Addition to the triple bond is slower because the vinyl radical products (with an sp^2 carbon) are less stable, and hence less readily formed, than alkyl radicals derived from alkenes.

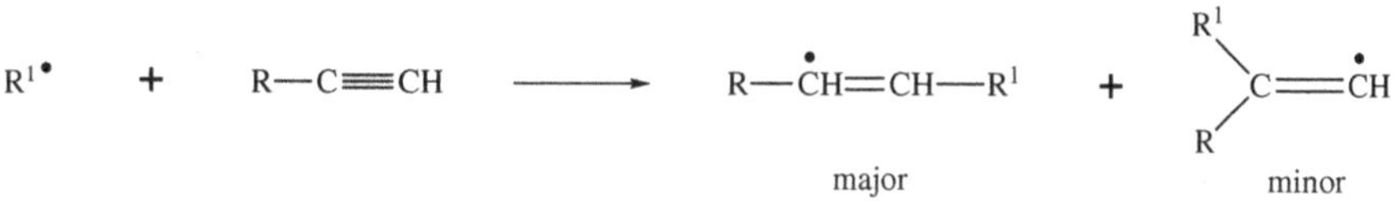

$$R^{1\bullet} \quad + \quad R-C\equiv CH \quad \longrightarrow \quad R-\overset{\bullet}{C}H=CH-R^1 \quad + \quad \underset{R}{\overset{R^1}{>}}C=\overset{\bullet}{C}H$$

major   minor

Fig. 4.10

Radicals, unlike cations or anions, can add to both electron-rich and electron-poor double bonds. The rate of addition does, however, depend on the size and polarity of the radical. Radicals with bulky side chains add slowly to substituted double bonds, nucleophilic radicals with electron-donating substituents add faster to electron-poor double bonds and electrophilic radicals (with electron-withdrawing substituents) add faster to electron-rich double bonds (see Section 4.7.5).

4.3.2 Intramolecular addition to alkenes and alkynes

Intramolecular radical additions to alkenes are very versatile and efficient reactions that lead to cyclic products. Cyclizations are generally faster than comparable intermolecular additions, and carbon-centred radicals can add to alkenes (intramolecularly) to form cyclic products containing a new carbon–carbon σ bond. These reactions are particularly useful for the formation of five- and six-membered rings. The additions are regioselective and radical addition usually occurs so as to form the smallest ring in an *exo* cyclization process (Fig. 4.11). This mode of cyclization leads to the formation of a radical located on an atom 'outside' (exocyclic to) the ring. The alternative *endo* cyclization produces a radical located on an atom 'inside' (endocyclic to), or part of, the larger ring. The *exo* cyclization produces a primary radical, whereas the *endo* cyclization generates a more stable secondary radical. Because the secondary radical is more stable, this is known as the thermodynamic product, whereas the primary radical is known as the kinetic product. The observed preference for the less stable (kinetic) cyclic product is surprising as intermolecular additions lead to the thermodynamic product (see Section 4.3.1). However, it has been shown that the p orbital of the radical can overlap more efficiently with the C=C π^* orbital in an *exo* transition state, so favouring addition to the 'internal' carbon of the double bond (see Section 4.7.4).

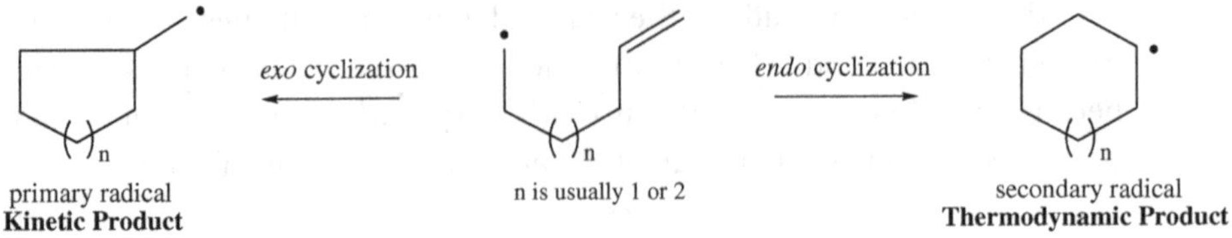

exo cyclization   endo cyclization

primary radical
Kinetic Product

n is usually 1 or 2

secondary radical
Thermodynamic Product

Fig. 4.11

Cyclization onto alkyne triple bonds also proceeds by the *exo* pathway to give the smaller ring. As for the intermolecular reactions, cyclizations are slower than for alkenes because a less stable vinyl radical is generated.

4.3.3 Addition to aromatics

The addition of carbon-centred radicals to aromatic compounds is slower than for addition to alkenes. This is because of the greater stability of the aromatic π system which is destroyed on addition of a radical. Reactive radicals (such as HO$^{\bullet}$ or Ph$^{\bullet}$) will, however, add to aromatic compounds, including benzene, to form cyclohexadienyl radicals that are stabilized by resonance (Fig. 4.12). These cyclohexadienyl radicals may react to regenerate the aromatic ring by expulsion of the initial radical (in a reversible

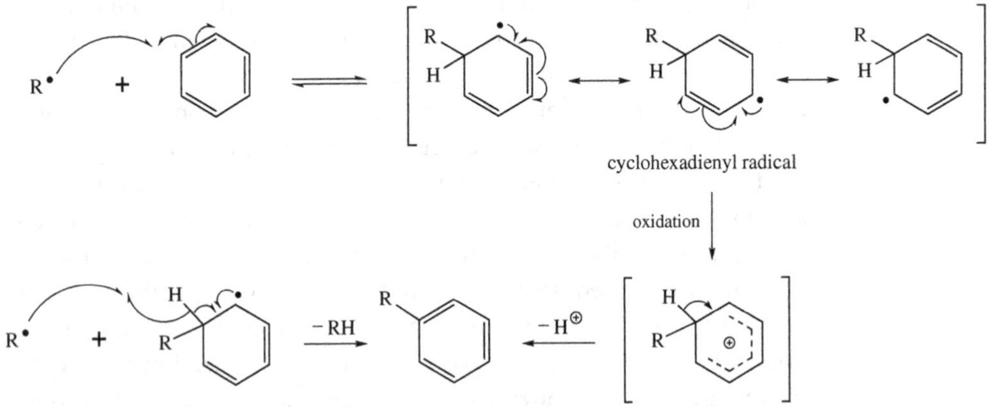

cyclohexadienyl radical

oxidation

Fig. 4.12

reaction) or by formally losing a hydrogen atom. The cyclohexadienyl radical is believed not to simply lose a hydrogen atom because the C—H bond is relatively strong (and, consequently, H• is not a good leaving group). The hydrogen atom may be abstracted by another radical or, alternatively, the radical could be oxidized (i.e. lose an electron) to form a cation which could then lose a proton.

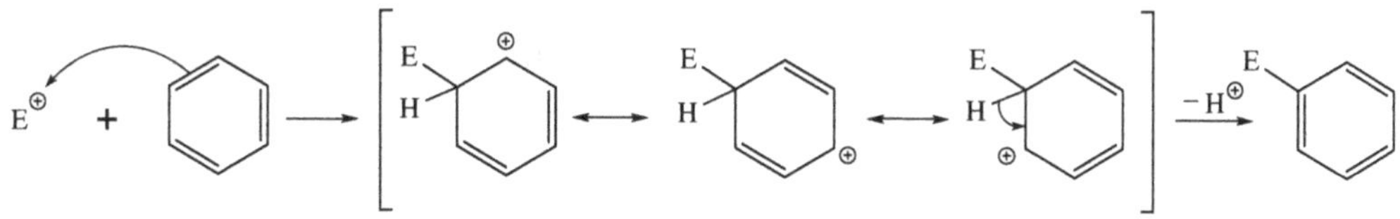

Fig. 4.13

Overall, this is a substitution reaction (called aromatic homolytic substitution) as a substituent (R) replaces a hydrogen atom on the benzene ring, and this is comparable to aromatic electrophilic substitution. Aromatic electrophilic substitution involves the addition of an electrophile to the electron-rich benzene ring to produce a positively charged Wheland intermediate (cf. the cyclohexadienyl radical) (Fig. 4.13).

The rate of radical addition depends on the nature of the radical and the benzene ring (cf. alkenes). Nucleophilic radicals add more quickly to aromatics containing electron-withdrawing groups, while electrophilic radicals add more quickly to aromatics containing electron-donating groups.

Additions to monosubstituted benzene rings are often regioselective and radicals generally prefer to attack the 2- and 4-positions (rather than the 3-position) of the ring. This can be explained by comparing the stability of the intermediate cyclohexadienyl radicals (Fig. 4.14). Attack at the 2- or 4-position will produce a more stable radical, because one canonical structure has the radical on a carbon atom adjacent to the substituent (Y) and this can stabilize the radical by conjugation or inductive effects. In contrast, reactions at the 3-position produce radicals that cannot be delocalized onto the substituent (Y) and are therefore less stable.

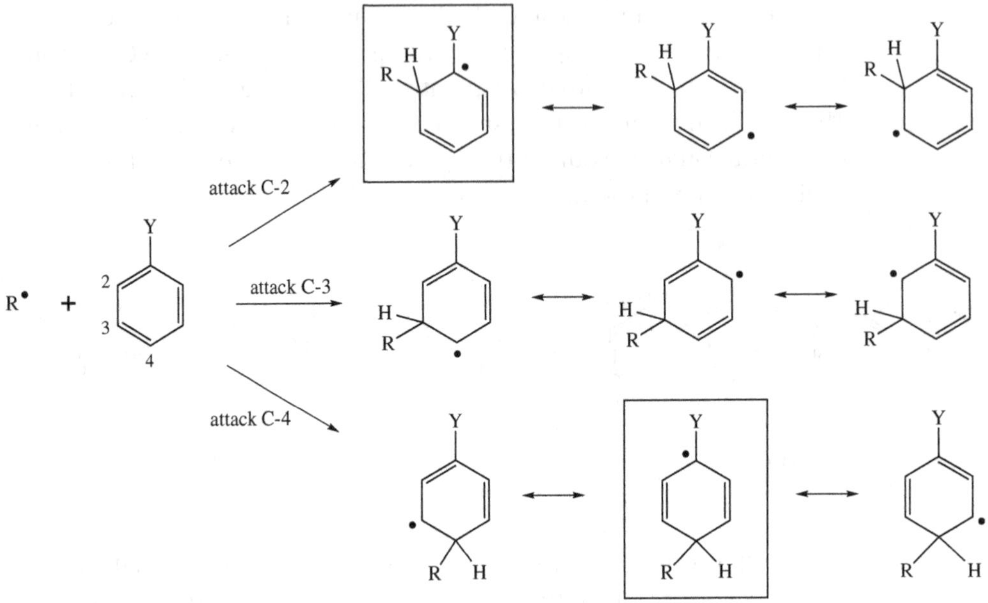

Fig. 4.14

Related intramolecular additions can be used to prepare five- and six-membered rings. Like the intermolecular reactions, these require the addition of a *reactive* radical to the benzene ring, otherwise the addition will be reversible and the starting radical will be eliminated (to regain aromaticity).

4.3.4 Addition to carbonyls

Whereas radicals readily add to alkenes, addition to carbonyls is usually less common because the C=O π bond is much stronger than the C=C π bond (by approximately 80 kJ mol^{-1}). Intramolecular reactions are possible using carbon-centred radicals which attack the carbonyl carbon (often reversibly) to generate alkoxyl radicals (in an *exo* cyclization) (Fig. 4.15). These reactions are similar to nucleophilic addition reactions (using, for example, cyanide), and the more nucleophilic the radical, the more likely it is that the attack will take place at the δ+ carbon. Metal-centred radicals that form strong σ bonds with oxygen will therefore behave as electrophilic radicals as they prefer to add (reversibly) to the δ– oxygen.

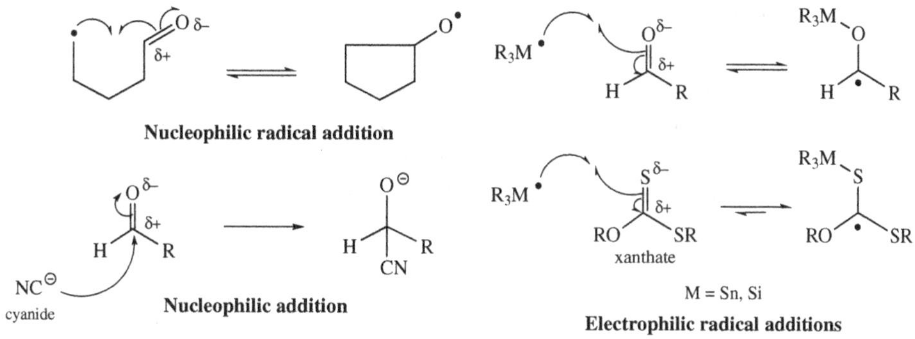

Fig. 4.15

Thiocarbonyls, in which the carbonyl oxygen is replaced by sulfur, undergo similar radical addition reactions. The C=S π bond is much weaker than the C=O π bond (by approximately 175 kJ mol⁻¹) and this ensures that radicals add more efficiently (and less reversibly) to thiocarbonyls than to carbonyls. A variety of functional groups containing the C=S bond (including xanthates) undergo radical additions, and reactions with metal-centred radicals (Fig. 4.15) are particularly important in functional group transformations (see Section 6.2.4).

Carbon-centred radicals can also add to the simplest carbonyl, namely carbon monoxide (CO), when the reaction is carried out under a high pressure of CO gas (typically 80 atm). Addition to the electron-deficient carbon atom of CO leads to the formation of an acyl radical, which can react further to produce, for example, an aldehyde (on abstraction of a hydrogen atom) (Fig. 4.16).

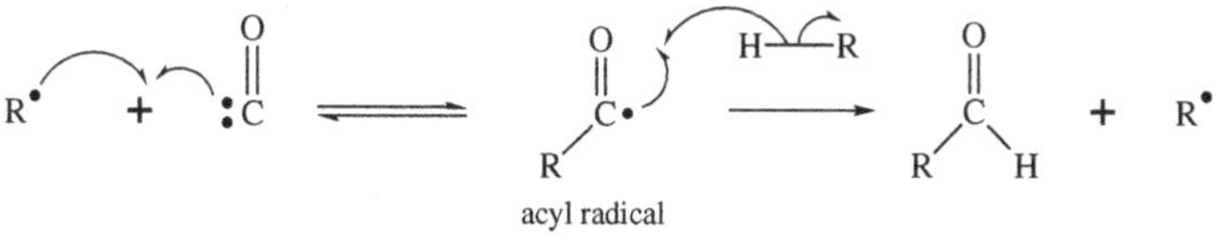

Fig. 4.16

4.3.5 Addition to oxygen

In Chapter 1, we saw that (in its ground state) molecular oxygen (O₂) exists as a diradical or triplet species as it contains two unpaired electrons. The two oxygen radical centres can undergo addition reactions so as to form two new covalent bonds, and carbon-centred radicals generally react rapidly to produce peroxyl radicals (Fig. 4.17). These reactions can be reversible, and the equilibrium depends on the stability of the carbon radical and the pressure of oxygen (the higher the pressure, the faster the addition). The peroxyl radicals can react further to form, for example, hydroperoxides and these reactions take place when organic compounds are gradually decomposed via autoxidation (see Section 6.2.3).

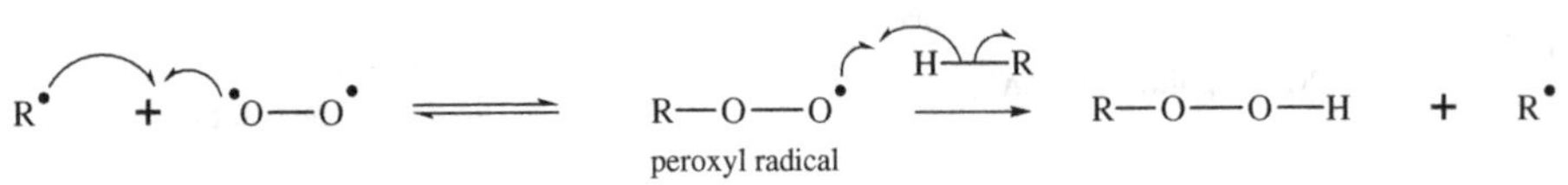

peroxyl radical

Fig. 4.17

4.4 Fragmentation reactions

Fragmentation reactions are the opposite of addition reactions. When a radical fragments by β-elimination, an unsaturated molecule, together with a new radical, is produced (Fig. 4.18). The driving force for these reactions is the increase in entropy, as two products are formed from one radical. For fragmentation to take place, the orbitals of the radical and the σ bond, which is cleaved, must lie in the same plane (see Section 4.7.4), and the mechanism is comparable to ionic eliminations proceeding by the E1 (unimolecular elimination) or E1cB (elimination from the conjugate base) pathways.

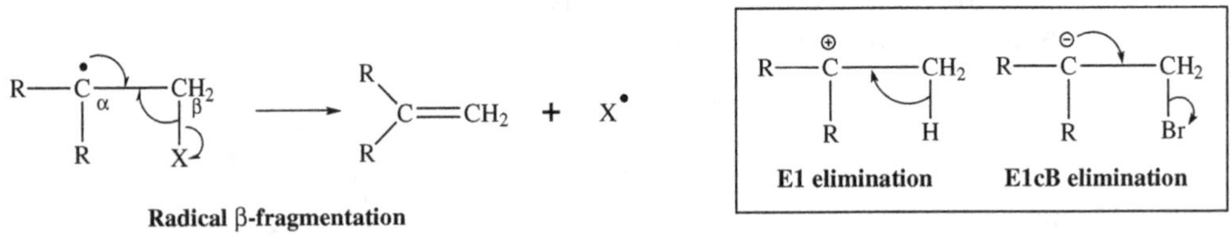

Radical β-fragmentation   **E1 elimination**   **E1cB elimination**

Fig. 4.18

Radicals can also fragment to lose small, stable molecules such as carbon dioxide (decarboxylation), carbon monoxide (decarbonylation) or nitrogen (N_2). Whereas fragmentations producing carbon dioxide and nitrogen are examples of β-elimination (as the alkyl group on the β-atom is eliminated), loss of carbon monoxide is an α-elimination process (Fig. 4.19).

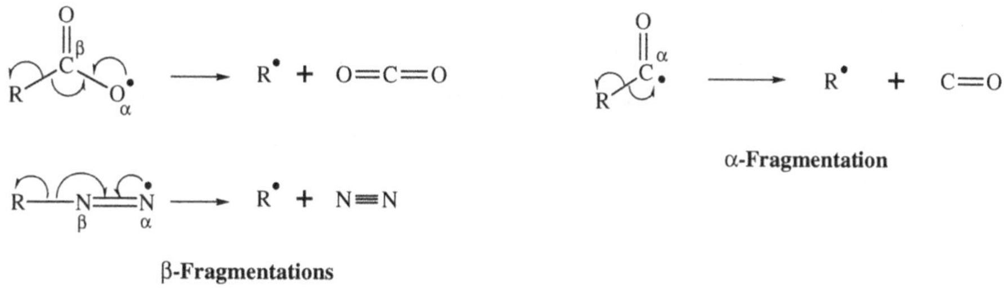

α-Fragmentation

β-Fragmentations

Fig. 4.19

4.4.1 Alkyl radicals

The rate of fragmentation of alkyl radicals to produce alkenes depends on the stability of the radical that is formed. As the alkene double bond is not particularly strong, an efficient fragmentation will require the elimination of a stable radical by cleavage of a weak σ bond (typically ≤ 290 kJ mol^{-1}). There are only a small number of radical 'leaving groups', including iodine and tin-centred radicals, that have sufficiently weak bonds to carbon to allow efficient fragmentation (Fig. 4.20).

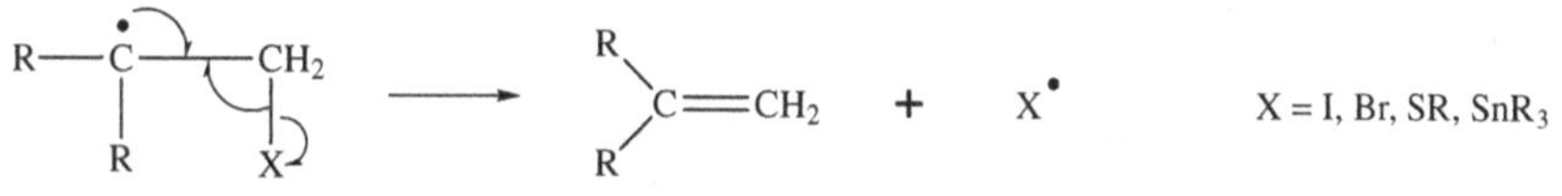

$X = I, Br, SR, SnR_3$

Fig. 4.20

4.4.2 Alkoxyl radicals

The fragmentation of alkoxyl radicals to produce carbonyls is favoured by the formation of a strong carbon–oxygen double bond (Fig. 4.21). Thus the *tert*-butoxyl radical ($^tBuO^•$) will fragment to form propanone (acetone) and a methyl radical even though the methyl radical is not particularly stable. With alkoxyl radicals bearing different alkyl (R) substituents, fragmentation leads predominantly to the most stable radical by breaking the weakest carbon–carbon bond.

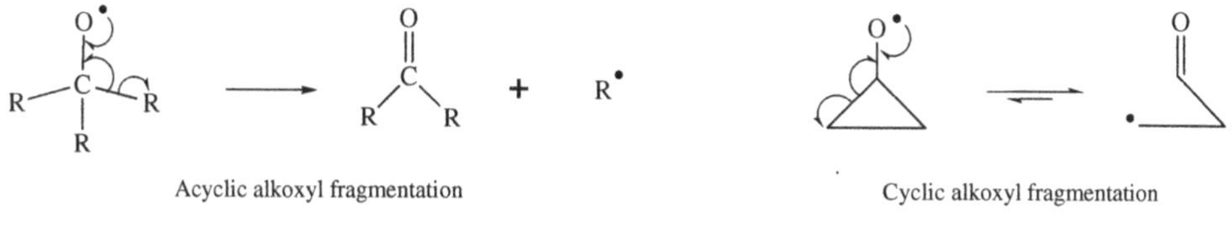

Acyclic alkoxyl fragmentation

Cyclic alkoxyl fragmentation

Fig. 4.21

73

Cyclic alkoxyl radicals can also fragment to give acyclic (aldehyde) radicals (Fig. 4.21), and the rate of fragmentation (or ring opening) depends on the size of the ring. The more strained (unstable) the ring, the faster the fragmentation, and three- and four-membered rings are particularly prone to ring opening.

4.4.3 Acyloxyl (or carboxyl), acyl and diazenyl radicals

Fragmentation of acyloxyl ($RCO_2^{\bullet}$) and diazenyl ($RN_2^{\bullet}$) radicals leads to the formation of alkyl or aryl radicals on elimination of carbon dioxide and nitrogen, respectively (Fig. 4.19). These are fast, irreversible reactions, and diazenyl radicals have particularly short lifetimes because fragmentation produces nitrogen, which is an excellent leaving group. Even fragmentation of Ph—N=N$^{\bullet}$ to give the phenyl radical, Ph$^{\bullet}$, is very fast (with a rate constant of around 10^8 s^{-1} at 40°C). In contrast, the decarbonylation of acyl radicals [$RC(\!\!=\!\!O)^{\bullet}$] is reversible (Fig. 4.19) because alkyl radicals can add to carbon monoxide (but not to carbon dioxide or nitrogen). This means that the rate of decarbonylation of acyl radicals is typically 100 times slower than the decarboxylation of acyloxyl radicals.

4.5 Rearrangements

Radical rearrangements involve the conversion of a precursor radical to a more stable product radical, and examples include intramolecular abstraction and addition (cyclization) reactions. Rearrangements are also observed when a radical undergoes an intramolecular addition (or cyclization) reaction, followed by a fragmentation (or ring opening) reaction. These reactions lead to the transfer of an (unsaturated) aryl, vinyl or carbonyl group, and the 1,2-shift (or transfer) of aromatic groups, for example, proceeds via cyclohexadienyl radicals (Fig. 4.22). These fragment so as to release ring strain and regain aromaticity, and the rate of the rearrangement depends on the substitution of both the radical and the benzene ring. Electrophilic radicals (with electron-withdrawing substituents), for example, will add more quickly to benzene rings with electron-donating substituents, while fragmentation to give tertiary radicals will be quicker than that to give secondary or primary radicals.

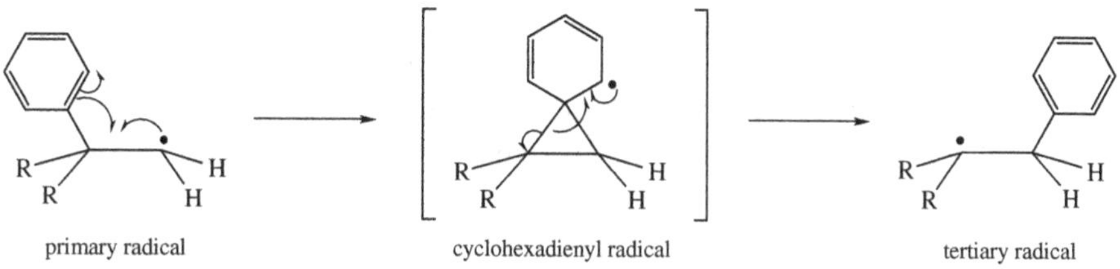

primary radical   cyclohexadienyl radical   tertiary radical

1,2-Aryl shift

Fig. 4.22

Whereas 1,2-aryl shifts (or neophyl rearrangements) are important in radical chemistry, the corresponding 1,2-alkyl (or hydrogen) shifts are extremely rare. Radical cyclization cannot take place on a saturated alkyl chain, and 1,2-alkyl rearrangements are usually only observed in carbocation reactions (e.g. the Wagner–Meerwein and Nametkin rearrangements).

4.6 Termination reactions

Radicals react to form non-radical products in termination reactions. This can involve the reaction of two radicals or, alternatively, the oxidation or reduction of a radical to a cation or anion, respectively.

4.6.1 Combination or coupling reactions

When two identical radicals (R$^•$) react, they combine to form a dimer (R—R) in a homocoupling or dimerization reaction. This reaction is the reverse of homolysis, and when two different radicals combine this is known as a heterocoupling reaction. These reactions, which produce a two-electron covalent bond, are usually very fast, essentially diffusion-controlled, processes. This means that almost every time two radicals collide, they will react with each other. Methyl radicals, for example, combine on every 4–6 collisions in the gaseous phase. However, the overall rate of the reaction can be low because, for efficient combination, we need a high concentration of the radicals; this is often difficult to achieve as most radicals have short lifetimes (because of reaction with other molecules). The best yields are therefore derived from stabilized radicals or from reactions that generate a high concentration of radicals.

Radical combination is observed when azo compounds (or peroxides) are thermally or photolytically decomposed in solution (Fig. 4.23). A solvent shell (or cage) surrounds the two radical products, together with nitrogen, and the radicals cannot escape by diffusion, particularly if the solvent is viscous. The radicals are held in close proximity as a 'geminate pair' (as shown by chemically induced dynamic nuclear polarization, CIDNP; see Section 2.2.2) and will readily combine, rather than undergo other reactions, to give a good yield of the 'cage product'. This is a rare example of a reaction that does not involve 'free' but 'caged' radicals. Even though most of the commonly used solvents (including petrol, benzene, methanol and water) have low viscosity, some cage recombination products are formed, and this decreases the proportion of 'free' radicals that can undergo other reactions.

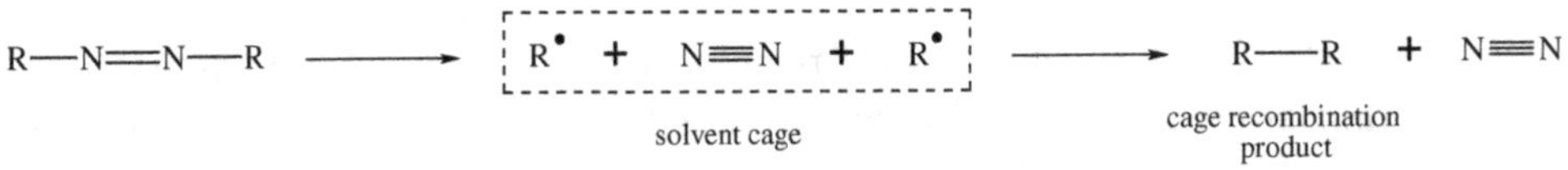

Fig. 4.23

4.6.2 Disproportionation

When two radicals react by disproportionation, one saturated product and one unsaturated product are formed (Fig. 4.24). The reaction mechanism involves the transfer of one β-hydrogen atom, and the driving force for these reactions is the formation of covalent bonds. For carbon-centred radicals, a new C=C π bond and a C—H σ bond are formed.

Disproportionation is in competition with radical combination, and these processes have similar (diffusion-controlled) reaction rates. For disproportionation, the radical must have a β-hydrogen, and the greater the number of β-hydrogens, the more likely it is that the radical will disproportionate. Steric factors are particularly important and bulky radicals, such as *tert*-butyl radicals [$^{\bullet}C(CH_3)_3$], tend to disproportionate more readily because the competing combination reaction requires reaction at *two* hindered carbon centres.

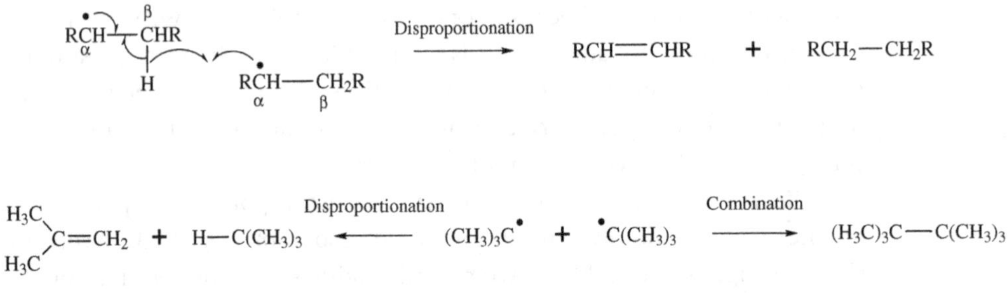

Fig. 4.24

4.6.3 Electron transfer reactions

Reduction or oxidation of a radical will lead to the formation of an anion or cation, respectively. The rate of these processes depends on the stability of the ion that is formed; for example, tertiary alkyl radicals are more easily oxidized to form carbocations (R_3C^+, stabilized by +I inductive effects) and primary radicals are more easily reduced to give carbanions (RCH_2^-). The functional groups adjacent to the radical play an important role and, whereas electron-donating substituents (e.g. OR, NR_2) facilitate radical oxidation, electron-withdrawing substituents (e.g. CO_2R, CHO, NO_2) facilitate radical reduction.

These redox reactions require (one-electron) oxidizing or reducing agents, and transition metal ions, which change their oxidation state by one, or electrolytic cells are most often employed. Oxidizing transition metals [such as silver(II) and cerium(IV)] prefer to gain an electron, while reducing metals [such as samarium(II)] prefer to lose an electron (Fig. 4.25).

Oxidation

$$R^\bullet - e^\ominus \longrightarrow R^\oplus$$

$$R^\bullet + M^n \longrightarrow R^\oplus + M^{n-1} \quad M^n = Ag(II), Ce(IV), Cu(II), Fe(III), Mn(III)$$

Reduction

$$R^\bullet + e^\ominus \longrightarrow R^\ominus$$

$$R^\bullet + M^n \longrightarrow R^\ominus + M^{n+1} \quad M^n = Cr(II), Sm(II), Ti(III), Fe(II), Cu(I)$$

Fig. 4.25

Although electron transfer leads to radical termination, the ionic products can undergo further reactions. Carbocations can react with a variety of nucleophiles, or lose a proton, while carbanions can be quenched with a proton or alternative electrophiles. Reduction using samarium(II), for example, leads to carbanions that can form a weak σ bond with the oxidized metal (e.g. R_3C-Sm^{III}), and these organometallic derivatives can undergo similar nucleophilic addition reactions to Grignard reagents (R—MgX). This radical–ionic reaction sequence is a very powerful method for preparing organic molecules in 'one-pot' reactions (see Section 5.4.1).

4.7 Reactivity and selectivity

We have seen (Section 4.6) that radicals can undergo a number of different and competitive reactions. These processes have different rates of reaction and if one reaction proceeds at a much faster rate than all the rest we have a selective and high-yielding process. Alternatively, if a variety of reactions proceed at similar rates, the radical will react unselectively to produce a number of different products. The rates of these reactions can vary enormously (Fig. 4.26) and, for example, the rate constants of abstraction reactions can vary by a factor of at least 10 000. The key factors that influence radical reactivity include enthalpy, entropy, steric effects, stereoelectronic effects, polarity and redox potential.

Combination or disproportionation
10^9 dm^3 mol^{-1} s^{-1}

Abstraction
(intermolecular)
10^4–10^8 dm^3 mol^{-1} s^{-1}

R$^•$

Addition
(intermolecular, to an alkene)
10^4–10^8 dm^3 mol^{-1} s^{-1}

Fragmentation
10^5–10^9 s^{-1}

Fig. 4.26

4.7.1 Enthalpy

Radical reactions will generally proceed so as to convert a radical into a more stable radical or non-radical product. Whereas combination reactions leading to non-radical products have similar (diffusion-controlled) reaction rates, the rates of radical/non-radical reactions can vary enormously (Fig. 4.26). As a guide, the more reactive the radical reactant and the more stable the product radical, the faster the rate of reaction. This explains why the reactive phenyl radical abstracts a chlorine atom much more quickly from carbon tetrachloride than does the *tert*-butyl radical (in solution at 25°C) (Fig. 4.27). Reaction with the phenyl radical is favoured because the carbon–chlorine bond in chlorobenzene is stronger than that in 2-chloro-2-methylpropane.

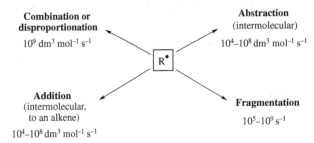

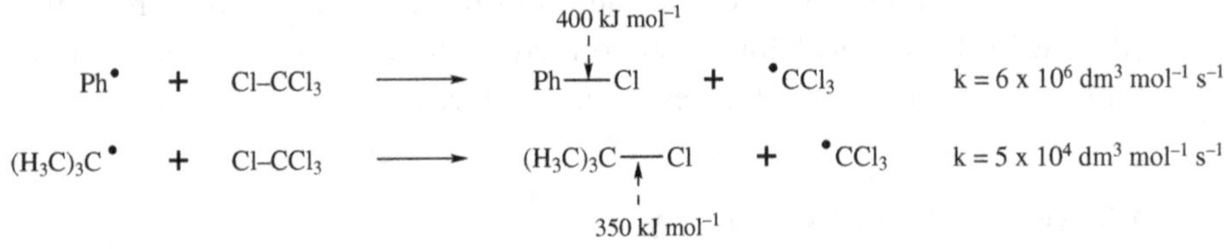

$$Ph^• \; + \; Cl\text{–}CCl_3 \longrightarrow Ph\overset{400 \text{ kJ mol}^{-1}}{\text{–}}Cl \; + \; {}^•CCl_3 \qquad k = 6 \times 10^6 \text{ dm}^3 \text{ mol}^{-1} \text{ s}^{-1}$$

$$(H_3C)_3C^• \; + \; Cl\text{–}CCl_3 \longrightarrow (H_3C)_3C\underset{350 \text{ kJ mol}^{-1}}{\text{–}}Cl \; + \; {}^•CCl_3 \qquad k = 5 \times 10^4 \text{ dm}^3 \text{ mol}^{-1} \text{ s}^{-1}$$

Fig. 4.27

We can therefore predict whether a radical reaction will take place by considering the energies of the bonds that are broken and those that are formed. This will provide an approximate enthalpy change ($\Delta H^{\ominus}$) for the reaction; if energy is released, the reaction is exothermic ($-\Delta H^{\ominus}$); if energy is absorbed, the reaction is endothermic ($+\Delta H^{\ominus}$) (Fig. 4.28). Exothermic reactions result in the formation of strong bonds and these can proceed rapidly (often spontaneously), whereas endothermic reactions (which lead to products with weaker bonds) are generally very slow.

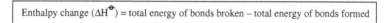

Enthalpy change ($\Delta H^{\ominus}$) = total energy of bonds broken – total energy of bonds formed

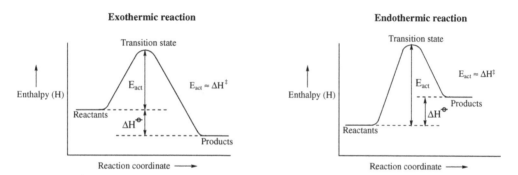

Fig. 4.28

However, this is not always the case, because the rate of reaction depends on the activation energy (E_{act}) which is approximately equal to the enthalpy of activation ($\Delta H^{\ddagger}$) (as $E_{act} = \Delta H^{\ddagger} + RT$ and, at 25°C, RT is only approximately 2.5 kJ mol^{-1}). This is a measure of the difference in bond energy between the starting materials and the *transition state*. However, we can use $\Delta H^{\ominus}$ as a guide to $\Delta H^{\ddagger}$ and, in general, the more exothermic (or more negative the value of $\Delta H^{\ominus}$ for) the reaction, the smaller $\Delta H^{\ddagger}$ and the faster the reaction. Enthalpies of activation for radical/non-radical reactions are usually small as most radicals are highly reactive, and so exothermic reactions can proceed very rapidly and even mildly endothermic reactions can proceed at a reasonable rate.

As an example, we will consider the addition of bromine and chlorine radicals to ethene (Fig. 4.29). The reactions lead to the formation of a carbon–bromine or carbon–chlorine bond at the expense of an alkene π bond. Both reactions are exothermic because the C—Br (285 kJ mol^{-1})

and C—Cl (340 kJ mol^{-1}) bonds are stronger than the C=C π bond (272 kJ mol^{-1}). However, we would expect the chlorine atom to react more quickly with the double bond because a stronger carbon–halogen bond is formed and so the reaction is more exothermic.

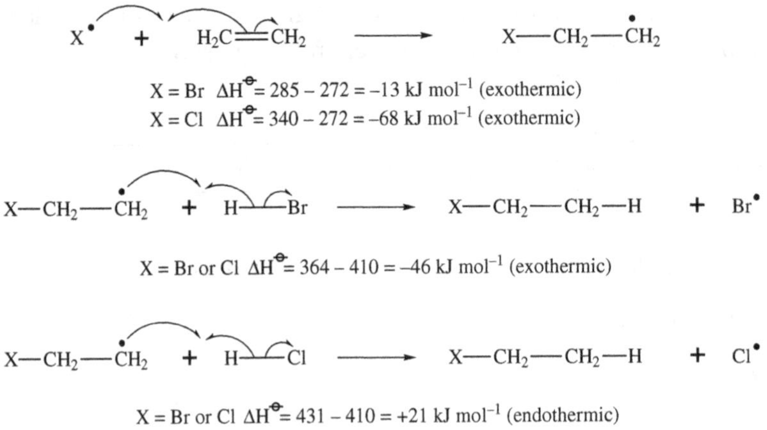

X = Br $\Delta H^{\ominus}$= 285 – 272 = –13 kJ mol^{-1} (exothermic)
X = Cl $\Delta H^{\ominus}$= 340 – 272 = –68 kJ mol^{-1} (exothermic)

X = Br or Cl $\Delta H^{\ominus}$= 364 – 410 = –46 kJ mol^{-1} (exothermic)

X = Br or Cl $\Delta H^{\ominus}$= 431 – 410 = +21 kJ mol^{-1} (endothermic)

Fig. 4.29

The primary radical products can react further and, if treated with a mixture of hydrogen chloride and hydrogen bromide, selective hydrogen abstraction from hydrogen bromide is likely to occur. This is because reaction with hydrogen bromide is exothermic; an H—Br bond (364 kJ mol^{-1}) is broken and a stronger C—H bond (410 kJ mol^{-1}) is formed. In comparison, the reaction with hydrogen chloride is endothermic because a strong H—Cl bond (431 kJ mol^{-1}) needs to be broken.

This analysis can be used to explain the selectivity observed on chlorination of alkanes. When a mixture of chlorine and propane is photolysed at 25°C, chlorine atoms are produced which can abstract hydrogen atoms to produce carbon-centred radicals and H—Cl (Fig. 4.30). The carbon-centred radicals then abstract a chlorine atom from molecular chlorine to give the primary chloride (5) in 43% yield and the secondary chloride (6) in 57% yield. Both reactions are exothermic as hydrogen atom abstraction produces an H—Cl bond ($\approx$ 430 kJ mol^{-1}) at the expense of a weaker primary C—H bond (410 kJ mol^{-1}) or secondary C—H bond (395 kJ mol^{-1}). On chlorine atom abstraction, a primary or secondary C—Cl bond ($\approx$ 410 kJ mol^{-1}) is formed and a weaker Cl—Cl bond (245 kJ mol^{-1}) is broken. The preferential formation of the secondary chloride (via a secondary radical) is therefore consistent with predominant cleavage of the weaker C—H bond. However, the selectivity is even more pronounced when we consider the number of hydrogen atoms in the starting material (two secondary and six primary hydrogen atoms) and the relative reactivity of the secondary and primary hydrogens (Fig. 4.30). By taking into account the statistical factor, the secondary C—H bond is shown to be around four times more likely to be broken than the primary C—H bond.

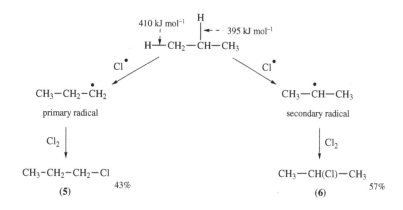

	yield of secondary chloride	57	
Relative reactivity of secondary to = primary radicals	number of secondary hydrogens	2	28.5
	=	=	= = 4
	yield of primary chloride	43	7.2
	number of primary hydrogens	6	

Fig. 4.30

We can represent the faster hydrogen atom abstraction from a secondary carbon in an energy diagram (Fig. 4.31). Hammond's postulate states that the transition states of exothermic reaction steps are generally reactant-like, whilst those of endothermic reaction steps are generally product-like. For this exothermic reaction, the (early) transition states of the hydrogen atom transfer steps are therefore closer in energy and geometry to the reactants. In the transition state, the radical character is spread from the chlorine to the carbon atom, and the secondary carbon will stabilize the radical more effectively than the primary carbon. The activation energy for reaction at the secondary carbon will therefore be lower and so the reaction giving the more stable secondary radical will be faster.

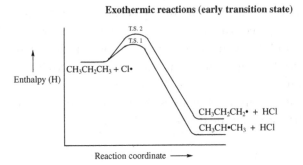
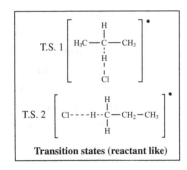

Exothermic reactions (early transition state)

T.S. 2

T.S. 1

Enthalpy (H)

CH₃CH₂CH₃ + Cl•

CH₃CH₂CH₂• + HCl

CH₃CH•CH₃ + HCl

Reaction coordinate ⟶

T.S. 1

T.S. 2

Transition states (reactant like)

Fig. 4.31

The same reaction using 2-methylpropane (at 25°C) gives the primary chloride **(7)** in 64% yield and the tertiary chloride **(8)** in 36% yield (Fig. 4.32). At first sight, it might appear that the stronger primary C—H bond is broken in preference to the tertiary C—H bond. However, taking into account the statistical factor (one tertiary to nine primary hydrogen atoms), it can be seen that the tertiary C—H bond is around five times more likely to be broken than the primary C—H bond. (You might like to try this calculation yourself.)

CH₃—C—CH₂—H X₂, hv ⟶ CH₃—C—CH₂—X + CH₃—C—CH₃

385 kJ mol⁻¹   410 kJ mol⁻¹

(7) X = Cl, 64%   (8) X = Cl, 36%
(9) X = Br, <1%   (10) X = Br, >99%

Fig. 4.32

Therefore, for chlorination reactions at 25°C, the relative reactivity of tertiary to secondary to primary C—H bonds is around 5 : 4 : 1, and this is consistent with the order of the bond dissociation energies. The ratio depends on the reaction temperature: the higher the temperature, the less selective the chlorination. At temperatures around 500–600°C, almost every collision will have enough energy to lead to a reaction and this produces a statistical distribution of products, the ratio being determined by the number of hydrogen atoms on each carbon.

The ratio also depends on the halogen, and bromination reactions can be highly selective. When a mixture of bromine and 2-methylpropane is photolysed at 150°C, the tertiary bromide **(10)** is formed in >99% yield and the primary bromide **(9)** in <1% yield (Fig. 4.32). When the statistical factor is taken into account, calculations show that the tertiary C—H bond is >890 (and typically 1700) times more likely to be broken than the primary C—H bond. So, why is bromination so regioselective?

For the bromination reaction, the formation of primary and tertiary carbon radicals is endothermic because a weaker H—Br bond ($\approx$ 366 kJ mol^{-1}) is formed when a primary (410 kJ mol^{-1}) or tertiary (385 kJ mol^{-1}) C—H bond is broken. Overall, however, the reaction is exothermic as the reaction of carbon-centred radicals with molecular bromine is very favourable; a stronger primary or tertiary C—Br bond ($\approx$ 290 kJ mol^{-1}) is formed at the expense of a very weak Br—Br bond (193 kJ mol^{-1}). We can represent hydrogen atom abstraction by the bromine radical in an energy diagram (Fig. 4.33). For this endothermic reaction, the late transition states (from Hammond's postulate) of the hydrogen atom transfer steps are therefore closer in energy and geometry to the products. The carbon atom undergoing hydrogen atom abstraction is more 'radical-like', and the activation energies for reaction at the tertiary and primary carbons will therefore reflect the different stabilities of the tertiary and primary radicals. The regioselectivity is much greater for bromination reactions because, in chlorination reactions, the early (reactant-like) transition states have undergone little bond breaking, and so the stability of the radicals (or C—H bond strengths) will not have as major an influence on the activation energies. This means that the difference in activation energies for hydrogen atom abstractions in bromination will be larger than in chlorination, leading to more selective reactions.

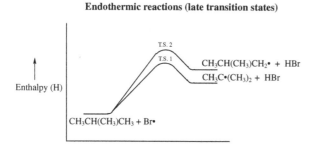

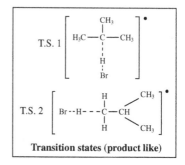

Fig. 4.33

These reactions have shown that the chlorine radical is more reactive (less selective) than the bromine radical. For the fluorine atom, we have an even more reactive radical, and this reacts violently with alkanes as the F—F bond is very weak (155 kJ mol^{-1}) and the H—F and C—F bonds are very strong (565 and 450 kJ mol⁻¹, respectively). Hydrogen atom abstraction from alkanes is therefore very exothermic, with an early transition state, and fluorination is not very selective. Thus, for example, reaction of fluorine with 2-methylpropane (at 25°C) gives a 6 : 1 ratio of primary to tertiary fluorides, and this is similar to the statistical product ratio of 9 : 1.

The iodine atom is the least reactive halogen radical and, like bromine, hydrogen atom abstraction from alkanes is endothermic because a weak H—I bond (297 kJ mol⁻¹) is formed and a stronger C—H bond (≈ 400 kJ mol⁻¹) is broken. Although the reaction of carbon radicals with molecular iodine is slightly exothermic [as a C—I bond (≈ 222 kJ mol⁻¹) is formed at the expense of an I—I bond (150 kJ mol⁻¹)], this is not sufficiently favourable for the overall reaction to be exothermic, and so iodine cannot react with alkanes.

4.7.2 Entropy

The enthalpy of a reaction gives only an approximate guide to selectivity in radical reactions. This is because the Gibbs free energy equation shows that temperature and entropy are also important factors in determining the outcome of a reaction (Fig. 4.34). For a reaction to be thermodynamically favoured, $\Delta G^\ominus$ should be negative; this occurs when there is a negative enthalpy change ($\Delta H^\ominus$), a positive entropy change ($\Delta S^\ominus$) and a high temperature (T). Therefore, reactions which produce an increase in entropy (or disorder) by increasing the number of molecular species on going from reactants to products are favoured. This explains why some endothermic reactions (positive $\Delta H^\ominus$) with an increase in entropy (positive $\Delta S^\ominus$) do not proceed spontaneously at room temperature. Higher temperatures are required so as to increase the $T\Delta S^\ominus$ contribution (above that of $\Delta H^\ominus$) to give a negative value of $\Delta G^\ominus$.

$$\boxed{\Delta G^\ominus = \Delta H^\ominus - T\Delta S^\ominus} \qquad -\Delta G^\ominus \text{ when } T\Delta S^\ominus > \Delta H^\ominus$$

$\Delta G^\ominus$ = Gibbs free energy; $\Delta H^\ominus$ = enthalpy change;
T = temperature; $\Delta S^\ominus$ = entropy change

Fig. 4.34

The decomposition of peroxides (RO—OR) and azo compounds (R—N=N—R) to form radicals is favoured by an increase in entropy. One peroxide molecule decomposes to give two alkoxyl (RO•) radicals, whereas an azoalkane will form three species: two carbon-centred radicals (R•) and nitrogen gas. The formation of a gas is particularly favoured because of the greater freedom of motion in gases (than in liquids and solids), resulting in an increase in disorder.

The driving force for a number of fragmentation reactions, which are often endothermic, is an increase in entropy. The decarbonylation of acyl radicals [RC(=O)•] is generally endothermic, but can proceed with reasonable rates (10^4–10^7 s^{-1} at 25°C) because two products, one of which is carbon monoxide gas, are formed. The increase in entropy also helps to explain why the decarboxylation of acyloxyl (or carboxyl) radicals can proceed so rapidly (10^6–10^{10} s^{-1} at 25°C). In addition to these reactions being exothermic, they are also favoured by entropy, and both of these factors contribute to give a fast and irreversible decarboxylation reaction.

For radical rearrangement reactions, a single reactant leads to a single product with nearly the same mass and an almost identical structure. Therefore, we can assume that $\Delta S^{\ominus}$ is approximately equal to zero, and hence $\Delta G^{\ominus} = \Delta H^{\ominus}$. Although cyclization reactions can be categorized as rearrangements, they generally show a decrease in (rotational) entropy, as bond rotation or the number of conformations in the cyclic product is not as great as that for the open-chain starting material.

The rate of a reaction depends on the entropy of activation ($\Delta S^{\ddagger}$), which measures the difference in entropy between the starting materials and the transition state. This is important for reactions that require a high degree of organization in the transition state, i.e. when reacting molecules must approach each other in a specific orientation. These reactions can be very slow, because forming the transition state requires a considerable loss of entropy. This is particularly important for the formation of large (more than nine-membered) rings. As the precursor chain length increases, the probability of the radical approaching and therefore reacting with the alkene (or similar group) decreases. Therefore, the longer the chain, the more conformations can be adopted, and the greater the loss of entropy ($\Delta S^{\ddagger}$) on formation of the cyclization transition state. The formation of smaller rings, particularly five- and six-membered rings, from shorter chains is therefore more favourable. For cyclizations giving three- and four-membered rings, the reactions are reversible because of ring strain, and the rates of ring opening (which lead to an increase in entropy) are higher than those for ring closure (Fig. 4.35).

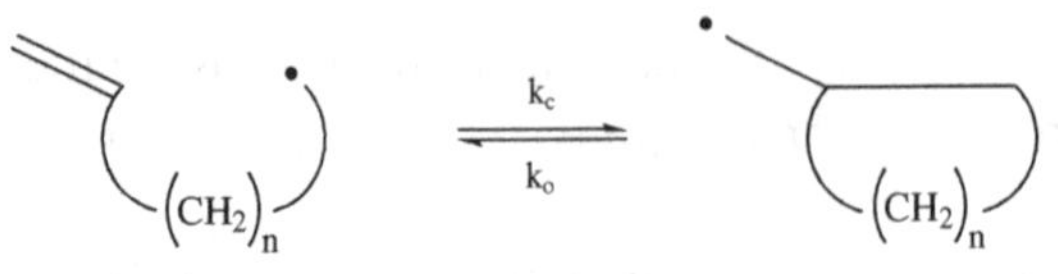

n	k_c/s^{-1} (25°C)	k_o/s^{-1} (25°C)
1	2×10^4	2×10^8
2	1×10^0	5×10^3
3	2×10^5	not observed
4	5×10^3	not observed
5	$<70 \times 10^0$	not observed

(k_c = rate of cyclization, k_o = rate of ring opening)

Fig. 4.35

Intramolecular hydrogen atom abstractions (S_Hi) are also controlled by entropy, and the linear geometry required in the transition state can be achieved in 1,5- and higher abstraction reactions. However, only 1,5- and 1,6-atom abstractions are commonly observed because, with longer chains, as for cyclization, there is a greater loss of entropy in the transition state.

A much greater loss of entropy occurs in intermolecular reactions when two reactants collide to form one product. Although the system has now become more ordered, for radical–radical reactions this does not have a major impact on the rate of reaction because combination and dispropor-tionation reactions (which lead to non-radical products) are very exothermic. A much more pronounced effect is observed for the reaction of radicals with non-radicals. These reactions are less exothermic (or enthalpy favoured) because a radical rather than a non-radical product is formed. Entropy now becomes much more important and the rates of these intermolecular reactions are considerably slower. This entropy factor explains why intermolecular radical addition reactions can be up to 10^5 times slower than related intramolecular cyclizations (under comparable reaction conditions).

4.7.3 Steric effects

A negative value for $\Delta G^\ominus$ tells us that a reaction can take place, but the rate of a reaction can be determined by the enthalpy of activation ($\Delta H^\ddagger$) (see Section 4.7.1). This is not only a measure of the difference in bond energy between the starting materials and the transition state, but also the differ-ence in bond strain. The more strained the transition state, the higher the value of $\Delta H^\ddagger$ and the slower the reaction. Therefore, even though two reactions can be thermodynamically favoured and have similar (negative) values for $\Delta H^\ominus$ they will not have the same rate if $\Delta H^\ddagger$ is different. We have seen that radical reactions usually have low values of $\Delta H^\ddagger$ because the majority of radicals are reactive. However, this is not the case for persistent or long-lived radicals that have very bulky substituents surrounding the radical centre. Reaction of these sterically hindered radicals would require a very strained transition state with a high enthalpy of activation and this is disfavoured.

Steric effects are used to explain the regioselective addition of radicals to alkenes (see Section 4.3.1). The radical preferentially attacks the less substituted (hindered) end of the double bond to give a less strained transition state with a lower $\Delta H^{\ddagger}$ in an anti-Markownikoff-type reaction. If we introduce substituents onto the double bond, the rate of addition is lowered because of greater steric interactions. Thus, for example, the methyl radical will add three times more quickly to ethene ($CH_2{=}CH_2$) than to the disubstituted alkene (*E*)-2-butene ($CH_3CH{=}CHCH_3$).

Steric effects can also explain why carbocyclic radicals (which are not able to undergo free rotation about the C—C bonds) can add to alkenes stereoselectively (see Section 7.4). The introduction of an adjacent chiral centre can make the two 'faces' of the radical non-equivalent, and this can lead to the alkene preferentially adding to the less hindered face (Fig. 4.36).

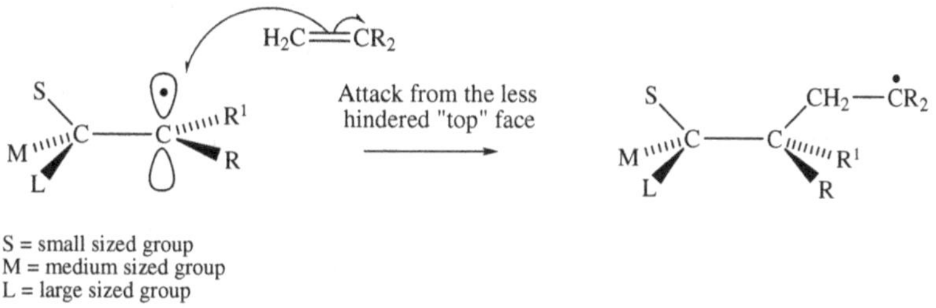
H2C=CR2
S
M
L
C
C
R1
R
Attack from the less
hindered "top" face
CH2—CR2
S = small sized group
M = medium sized group
L = large sized group

Fig. 4.36

4.7.4 Stereoelectronic effects

In Section 2.4, we saw that radicals with electron-rich or electron-poor substituents can be stabilized by interaction of the singly occupied orbital with an n, π or σ orbital. For effective stabilization, the interacting orbitals must overlap efficiently, and this will depend on their geometry, or position, in space. Similarly, for a radical to react, the singly occupied orbital must be able to overlap with either another of its own orbitals (for intramolecular reactions) or another radical or non-radical orbital in a different molecule. For reaction of a radical with a different molecule, there is often no restriction in the orbital geometries and they can rotate freely so as to combine with the maximum overlap.

For an S_H2 reaction, a radical and non-radical can therefore orientate themselves to give a linear transition state which maximizes the interaction of the radical orbital and the vacant σ* orbital of the bond which is broken (Fig. 4.37). This may, however, be difficult for very bulky molecules, where steric hindrance (see Section 4.7.3) can prevent the orbitals from becoming close enough for efficient overlap.

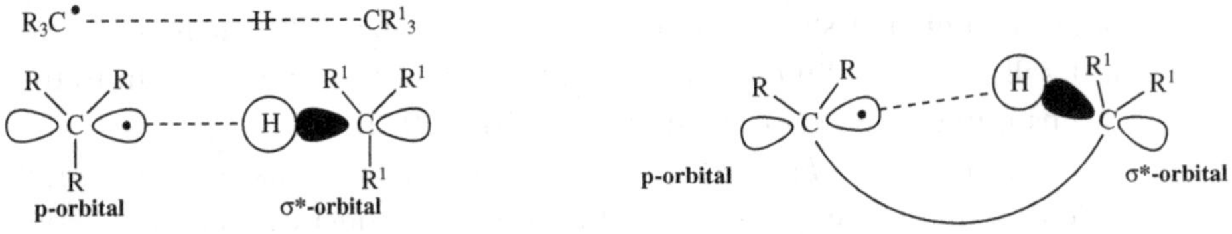

Intermolecular "linear" interaction

Intramolecular "approximately linear" interaction

Fig. 4.37

For intramolecular radical reactions, there is often a restriction in the orbital geometries. The structure of the starting material dictates how far the orbitals are apart and their relative position; if they are situated a long way apart or held rigidly on different sides of a ring, they will find it hard to interact. When the radical orbital is not in close proximity to the C—H σ* orbital, intermolecular hydrogen atom abstraction can compete with the intramolecular process. Even when the orbitals are very close together, they may not interact because their geometry can prevent orbital overlap. This so-called electronic strain explains why 1,3- and 1,4-hydrogen atom transfers are very slow. The short chain length restricts the orbital positions and so the radical orbital cannot attack the vacant σ* orbital at an angle of approximately 180° to give the most efficient overlap.

Stereoelectronic factors have been used to explain why the 5-*exo* mode of cyclization is faster than the competitive 6-*endo* cyclization. For the hex-5-en-1-yl radical, the rate of 5-*exo* cyclization is around 60 times faster than that for the 6-*endo* reaction at 25°C (Fig. 4.38). This is surprising as the 6-*endo* product is a secondary radical and (thermodynamically) more stable than the primary radical formed on 5-*exo* cyclization. In addition, the five-membered cyclopentane ring is more strained than the cyclohexane ring. The cyclization is therefore under kinetic control, and this is because the singly occupied orbital of the radical attacks the alkene at an angle close to 107° and so can overlap more favourably with the alkene π* orbital at the 'internal' carbon atom. The carbon chain will position the radical's p orbital closer to the internal (rather than the external) alkene carbon, and energy calculations have confirmed that the smaller ring is formed because the chair-like transition states are less strained and lower in energy.

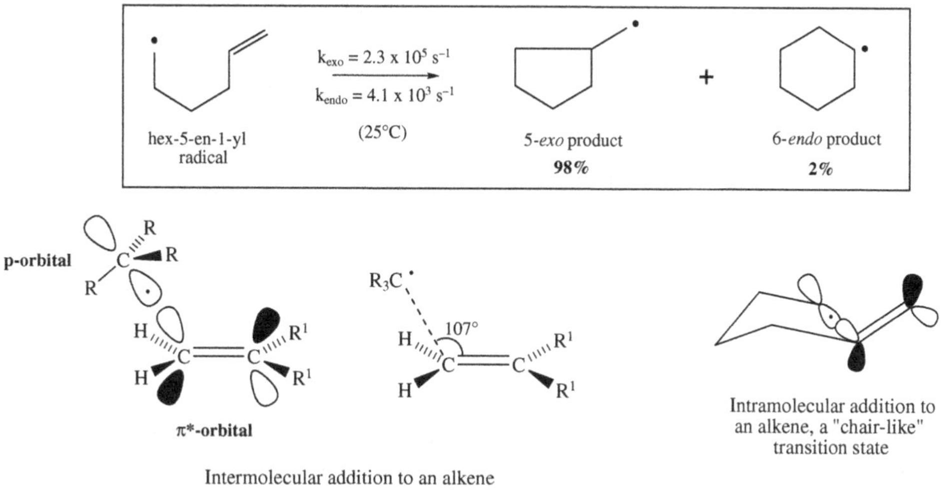

hex-5-en-1-yl radical

$k_{exo} = 2.3 \times 10^5 \text{ s}^{-1}$

$k_{endo} = 4.1 \times 10^3 \text{ s}^{-1}$

(25°C)

5-*exo* product
98%

+

6-*endo* product
2%

p-orbital

π*-orbital

Intermolecular addition to an alkene

Intramolecular addition to an alkene, a "chair-like" transition state

Fig. 4.38

Radical fragmentation reactions are also under stereoelectronic control. For β-elimination or fragmentation of a carbon-centred radical, the radical p orbital and the σ orbital of the C—X bond that is broken must lie in the same plane so that the double bond can be easily formed (Fig. 4.39). This is not a problem for molecules that can rotate freely around the central carbon–carbon bond, as rotation can position the radical and C—X bond in the correct orientation. The cyclopropylmethyl radical, for example, can undergo rotation around the exocyclic carbon–carbon bond to align the p and σ orbitals, and very rapid β-elimination occurs to open the strained three-membered ring. In contrast, the cyclopropyl radical cannot undergo ring opening because the carbon–carbon bonds within the ring cannot rotate, and so the p and σ orbitals lie at 90° (or orthogonal) to one another. This is in spite of the fact that the three-membered ring is very strained and ring opening is thermodynamically favoured.

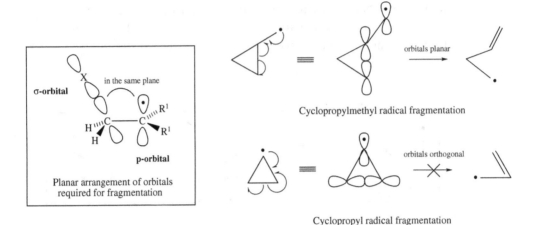

Cyclopropylmethyl radical fragmentation

Cyclopropyl radical fragmentation

Fig. 4.39

4.7.5 Polarity

For reactions to occur, the interacting or frontier orbitals must not only overlap efficiently, but also have similar energies. The occupied frontier orbital for a radical is called the singly occupied molecular orbital (SOMO), and in Section 2.4.3 we saw that radicals bearing different substituents have different SOMO energies. Radicals adjacent to electron-donating groups interact with a filled orbital to give a high-energy SOMO, while radicals next to electron-withdrawing groups interact with an unfilled orbital to give a low-energy SOMO. These SOMO energies lie somewhere between the highest occupied molecular orbital (HOMO) and lowest unoccupied molecular orbital (LUMO) of non-radicals. Therefore, for the reaction of a radical with a non-radical, we need to consider the SOMO–HOMO and SOMO–LUMO interactions (Fig. 4.40a,b). In both cases, the interaction will lead to a decrease in energy and the formation of a bond; a SOMO–HOMO interaction places two of the three electrons in a low-energy bonding orbital, whereas a SOMO–LUMO interaction places the single electron in a low-energy bonding orbital. The energy of the SOMO will determine whether interaction with the HOMO or (higher energy) LUMO predominates. Electrophilic radicals (with a low-energy SOMO) will be closer in energy to the HOMO, and therefore the SOMO–HOMO interaction will predominate (Fig. 4.40a). In comparison, nucleophilic radicals (with a high-energy SOMO) will be closer in energy to the LUMO, and therefore the SOMO–LUMO interaction will predominate (Fig. 4.40b).

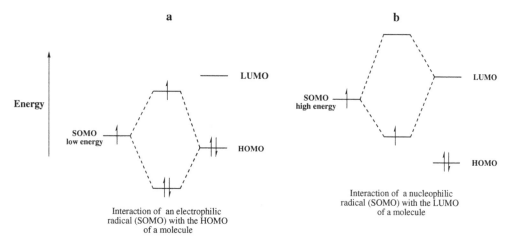

Fig. 4.40

These orbital interactions can explain the selectivity observed when the carbon–hydrogen bond is cleaved in hydrogen atom abstraction reactions. The energies of the HOMO (σ) and LUMO (σ^*) orbitals of the C—H bond depend on the substituents attached to the carbon. Electron-withdrawing substituents will lower the HOMO and LUMO energies, whereas electron-donating substituents will raise these energies (as for radical orbitals). Therefore, in propanoic acid, the methylene (CH_2) carbon–hydrogen bond orbitals are lower in energy than those of the methyl (CH_3) carbon–hydrogen bonds because they are adjacent to the (electron-withdrawing) carboxylic acid group (Fig. 4.41). When an electrophilic chlorine radical reacts with propanoic acid, a methyl hydrogen is selectively abstracted because the low-energy SOMO will interact with the higher energy carbon–hydrogen HOMO. In other words, the higher the HOMO energy, the greater the interaction with a low-energy SOMO. Alternatively, when a nucleophilic methyl radical reacts with propanoic acid, a methylene hydrogen is selectively abstracted. This results from interaction of the high-energy SOMO with the lower energy C—H LUMO; the lower the LUMO energy, the greater the interaction with a high-energy SOMO.

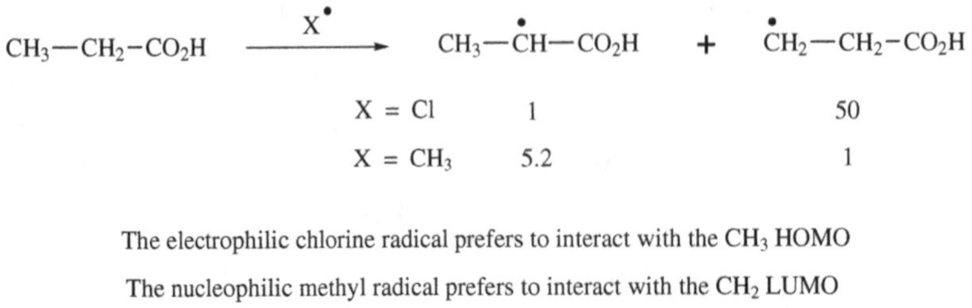

$$CH_3—CH_2—CO_2H \xrightarrow{\quad X^\bullet \quad} CH_3—\overset{\bullet}{C}H—CO_2H \quad + \quad \overset{\bullet}{C}H_2—CH_2—CO_2H$$

	X = Cl	1	50
	X = CH$_3$	5.2	1

The electrophilic chlorine radical prefers to interact with the CH$_3$ HOMO

The nucleophilic methyl radical prefers to interact with the CH$_2$ LUMO

Fig. 4.41

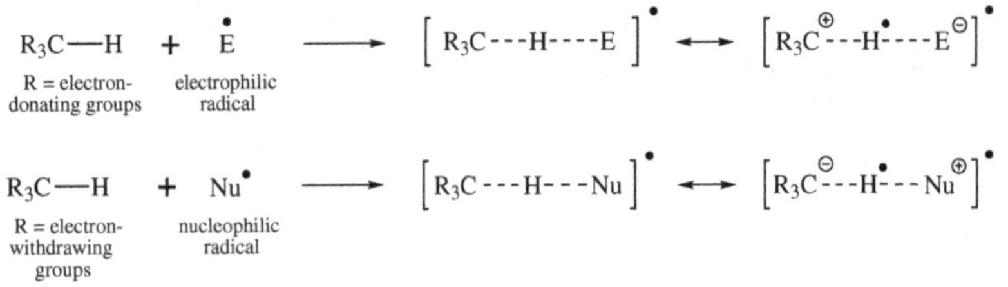

$$R_3C—H \quad + \quad \overset{\bullet}{E} \longrightarrow \left[R_3C\text{---}H\text{----}E \right]^\bullet \longleftrightarrow \left[R_3\overset{\oplus}{C}\text{---}\overset{\bullet}{H}\text{----}E^\ominus \right]^\bullet$$

R = electron-donating groups   electrophilic radical

$$R_3C—H \quad + \quad Nu^\bullet \longrightarrow \left[R_3C\text{---}H\text{---}Nu \right]^\bullet \longleftrightarrow \left[R_3\overset{\ominus}{C}\text{---}\overset{\bullet}{H}\text{---}\overset{\oplus}{Nu} \right]^\bullet$$

R = electron-withdrawing groups   nucleophilic radical

Fig. 4.42

These effects can also be understood in terms of partial charge separa-tion in the transition state. If an electrophilic radical (E$^\bullet$) reacts with R$_3$C—H, we can represent the transition state by a resonance form in which the radical has accepted an electron to become E$^-$ (Fig. 4.42). As the radical is electrophilic, it can readily accept an electron from the C—H bond, and the resulting radical cation can break down to give R$_3$C$^+$ and H$^\bullet$. The carboca-tion, R$_3$C$^+$, is stabilized by electron-donating groups (R), and therefore the more electron rich the carbon–hydrogen bond, the more likely it is to react with an electrophilic radical. For reaction of the chlorine radical with propanoic acid (Fig. 4.41), we would therefore expect the electrophilic radical to react preferentially with the methyl C—H bond because this has the highest electron density. Alternatively, reaction of R$_3$C—H with a nucleophilic radical (Nu$^\bullet$) will produce a transition state in which a resonance structure contains the carbanion R$_3$C$^-$. This is stabilized by electron-withdrawing groups (R), and therefore the more electron poor the carbon–hydrogen bond, the more likely it will be to react with a nucleophilic radical. Hence, the nucleophilic methyl radical preferentially attacks a methylene carbon–hydrogen bond in propanoic acid because the acid substituent reduces the electron density (and so stabilizes the carban-ion in the transition state).

We can extend the frontier molecular orbital approach to the reaction of radicals with alkenes. These addition reactions involve the interaction of the radical SOMO with the HOMO (π) and LUMO (π^*) of the π double bond. Nucleophilic radicals (with high-energy SOMOs) will prefer to react with alkenes containing electron-withdrawing substituents as these have low-energy LUMOs. The nucleophilic cyclohexyl radical ($C_6H_{11}{}^\bullet$), for example, adds 8500 times more rapidly to acrolein ($CH_2{=}CH{-}CHO$) than to 1-hexene ($CH_2{=}CH{-}C_4H_9$) at 20°C. Conversely, electrophilic radicals (with low-energy SOMOs) will prefer to react with alkenes containing electron-donating substituents as these have high-energy HOMOs.

This explains the alternating sequence of monomers that is observed on copolymerization of electron-rich and electron-poor alkenes (Fig. 4.43). Radical addition to an electron-rich alkene, such as styrene, will generate a nucleophilic (benzylic) radical. This would prefer to react with an electron-poor alkene; therefore, reaction with, for example, methyl acrylate is faster than reaction with another molecule of styrene. This produces an electrophilic radical (adjacent to the electron-withdrawing ester) that prefers to react with the electron-rich double bond of styrene, and this process is repeated to produce the alternating polymer in a remarkably selective reaction.

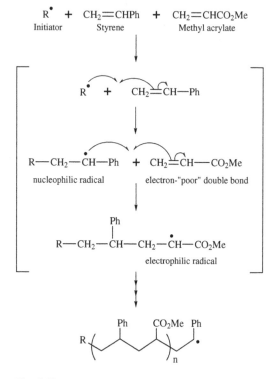

Fig. 4.43

4.7.6 Redox potential

When electron transfer methods (e.g. transition metals) are used to prepare radicals, the product radicals may be oxidized or reduced to cations or
anions, respectively, before they have the chance to undergo any radical
reactions. The rate of oxidation or reduction of radicals will be determined
by the redox potential of the radical and oxidant or reductant. Nucleophilic radicals, with electron-donating substituents, will be more easily
oxidized to form cations, whereas electrophilic radicals, with electron-
withdrawing substituents, will be more easily reduced to form anions (see
Section 4.7.5).

   Therefore, if a nucleophilic radical is generated in the presence of an
oxidizing agent, the radical may be oxidized so quickly that addition to an
electron-poor double bond cannot take place. However, an electrophilic
radical, such as the malonyl radical, would be expected to undergo very
slow oxidation and this radical would add to an electron-rich (styrene)
double bond before oxidation takes place (Fig. 4.44). This produces a
nucleophilic radical that could then undergo oxidation to form a cation.
These are useful reactions as they combine radical and ionic intermediates;
for efficient reactions, we need to be able to either selectively oxidize or
reduce the product radical (in the presence of the first-formed radical).
This works best when the reactant and product radicals have different
polarities, and therefore also have different substitution patterns.

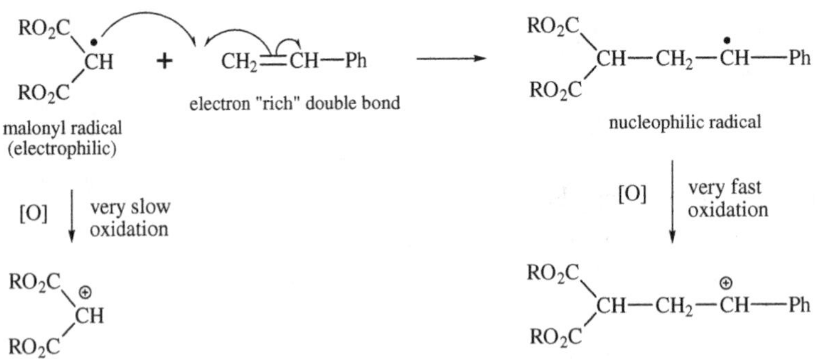

Fig. 4.44

4.8 Summary

Radicals can undergo a variety of different propagation and termination
reactions to form chemical bonds. These reactions, producing lower
energy non-radical products, can be very selective, and we can predict the
formation of the predominant or exclusive product from an understanding

of the factors that control the reactivity of radicals. The most important factor is enthalpy, and the outcome of a reaction can often simply be related to the formation of a strong bond at the expense of a weak bond. Thus, reactive (high-energy) radicals, such as the phenyl radical, will tend to react less selectively than more stable radicals (e.g. the benzyl radical) to form a much stronger C—H or C—C bond. Entropic and steric effects are also important, particularly when using reactants with bulky substituents, and the structure of the radical also influences the geometry and energy of the SOMO. Stereoelectronic factors can explain why some radical reactions are under kinetic control and proceed to give the thermodynamically less stable product, while a frontier molecular orbital approach can explain why electrophilic and nucleophilic radicals can react at different rates with the same substrate.

Further reading

Beckwith, A.L.J. (1981) Regio-selectivity and stereo-selectivity in radical reactions. *Tetrahedron*, **37**, 3073–3100.

Beckwith, A.L.J. (1993) The pursuit of selectivity in radical reactions. *Chemical Society Reviews*, **22**, 143–151.

Giese, B. (1983) Formation of CC bonds by addition of free radicals to alkenes. *Angewandte Chemie, International Edition in English*, **22**, 753–764.

Tedder, J.M. (1982) The importance of polarity, bond strength and steric effects in determining the site of attack and the rate of free radical substitution in aliphatic compounds. *Tetrahedron*, **38**, 313–329.

Walton, J.C. (1998) Homolytic substitution: a molecular ménage à trois. *Accounts of Chemical Research*, **31**, 99–107.

CHAPTER 5

Radicals in Synthesis

5.1 Introduction

The wide diversity and range of ionic reactions begs the following questions. Why use radical reactions in synthesis? What are the advantages of using radicals over anion or cation intermediates? This chapter aims to address these questions, and starts by discussing some important comparisons between the use of radicals and ions in synthesis. Aspects of chemo-, regio- and stereoselectivity are highlighted, together with polarity and practical considerations. The importance of chain and non-chain radical reactions is then discussed, and these reactions are illustrated by reference to some radical reagents that are commonly used in synthesis.

5.2 Radicals vs. ions

5.2.1 Reaction conditions

Whereas ionic reactions generally proceed at high or low pH, radical reactions proceed under mild and neutral conditions. The main disadvantage of the acid or alkaline conditions associated with ionic reactions is that acid- or base-sensitive substrates can be decomposed and chiral centres can be racemized. These substrates can, however, undergo radical transformations without fear of decomposition or racemization because of the milder reaction conditions.

For ionic transformations, changing the solvent can alter the solvation and hence the reactivity of cations and anions, as a solvent shell can protect the ions and slow down their reaction. A bulky solvent shell can therefore lower the reactivity of the ion, and steric interactions can prevent the desired reaction. As a result, the reaction yields will be affected and a range of solvents may have to be investigated for optimum results. Solvent effects, however, are much less important in radical reactions as radicals are uncharged and usually have little interaction with the solvent. This means that changing the reaction solvent will be less important, and the yields of products can be similar when using solvents of different polarity. In addition, radical reactions can often be used to assemble very hindered molecules that cannot be prepared using ionic intermediates.

For radical reactions, solvents with strong bonds are often used to ensure that there is no reaction with the radical intermediates. This is particularly important if very reactive radicals are involved; the phenyl radical is usually prepared in benzene as hydrogen atom abstraction from the solvent is slow (as this would regenerate the phenyl radical and give benzene). Toluene is often used in place of benzene as the reaction solvent (as benzene is a carcinogen), although toluene can inhibit some chain reactions involving reactive radicals, which can add to the aromatic nucleus or abstract a hydrogen atom from the methyl group. One of the best solvents for radical reactions is water, the O—H bond being stronger than, for example, C—H or C—halogen bonds present in organic reactants. For ionic reactions, this is usually the worst possible solvent and anhydrous reaction conditions (e.g. 'dry' solvents) are often employed so as to avoid hydrolysis or protonation of cations or anions, respectively.

We should also be aware of the role of oxygen in radical chemistry (see Section 4.3.5). If a carbon-centred radical ($R^•$) is generated in the presence of oxygen, peroxyl radicals ($ROO^•$) are very rapidly formed. The reaction is only reversible with exceptionally stable carbon radicals, such as $Ph_3C^•$, and although the equilibrium depends on the amount of oxygen present, for most aliphatic radicals the reverse reaction is not observed. Peroxyl radicals can react further to form, for example, hydroperoxides ($ROOH$) and, if these products are not desired, reactions should be conducted under an oxygen-free atmosphere (e.g. nitrogen or argon). Similar conditions are also employed in organometallic chemistry because, for example, organolithium (RLi) and Grignard reagents ($RMgX$) can react with oxygen to form hydroperoxides. The hydroperoxides are likely to be formed from carbon-centred radicals which have been shown to be present when Grignard reagents are prepared from magnesium and alkyl halides (Fig. 5.1).

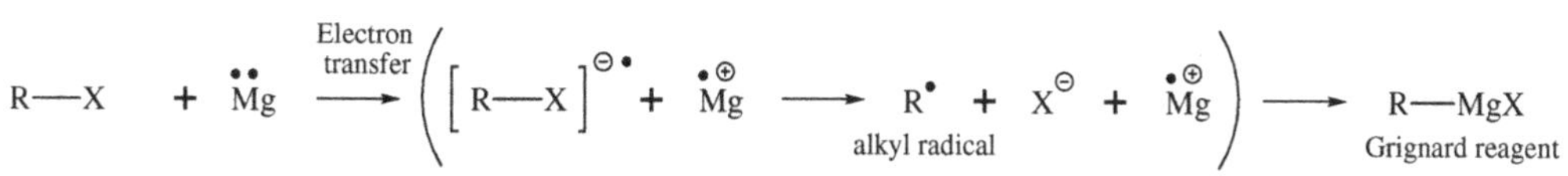

$$R\text{—}X \quad + \quad \overset{\bullet\bullet}{Mg} \quad \xrightarrow{\substack{\text{Electron}\\ \text{transfer}}} \quad \left(\left[R\text{—}X \right]^{\ominus \bullet} + \overset{\bullet\oplus}{Mg} \quad \longrightarrow \quad \underset{\text{alkyl radical}}{R^\bullet} \; + \; X^\ominus \; + \; \overset{\bullet\oplus}{Mg} \right) \quad \longrightarrow \quad \underset{\text{Grignard reagent}}{R\text{—}MgX}$$

Fig. 5.1

5.2.2 Radical concentration

Radical reactions are also influenced by concentration. The lower the concentration of radicals in solution, the more likely they are to react with the solvent; at high dilution, even alkyl radicals can add to benzene. For

combination (coupling) or disproportionation reactions, this is not a problem. A concentrated solution of the precursor can be used to produce a
high concentration of radicals, which can then combine or disproportionate at diffusion-controlled rates (Fig. 5.2). However, propagation reactions, in which a radical reacts with a non-radical, such as an alkene, are
more problematic because of the slower rates of reaction. At high concentrations, the radicals will prefer to undergo faster combination reactions to
produce a more stable non-radical (rather than radical) product. A reason-
ably dilute solution (typically 10^{-7}–10^{-8} mol dm^{-3}) of the radical precursor
and alkene is generally employed, and the yield of the propagation reaction can be maximized in two ways: firstly, by using an excess of the
alkene to increase the likelihood of reaction with the radical (R$^\bullet$); and,
secondly, by slow addition of a radical initiator; this generates a low concentration of radicals from reaction with the radical precursor (R—X).

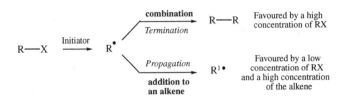

Fig. 5.2

In contrast, concentration is less important for ionic reactions. High precursor concentrations can be used and ions can be generated without fear
of coupling as two cations (or anions) will be electrostatically repelled, not
attracted!

5.2.3 Chemoselectivity

To transform a complex precursor with many functional groups into a target molecule, we often require one functional group to react in preference
to the others. For this, we need to employ chemoselective reactions. For
radical reactions, the bond strengths will serve as a guide to the chemo-
selectivity. If one bond in the precursor is much weaker than all the others,
then we might expect this to react selectively. For example, compounds
containing both C—Br and C—Cl bonds can react to selectively cleave
the weaker C—Br bond. Functional groups with strong σ or π bonds are
therefore inert under radical conditions. This includes amines, alcohols,

amides and carboxylic acids that have very strong N—H or O—H bonds (≈ 400 kJ mol^{-1}). These functional groups generally need to be 'protected' when ionic transformations are carried out elsewhere in the molecule; a protecting group is introduced so as to avoid side reactions involving deprotonation or protonation of these groups (Fig. 5.3). However, these bonds are sufficiently strong that homolytic cleavage is very rare. This means that protection of such groups is generally not necessary, which avoids two cumbersome steps (protection and deprotection) and leads to a quicker approach to the target molecule.

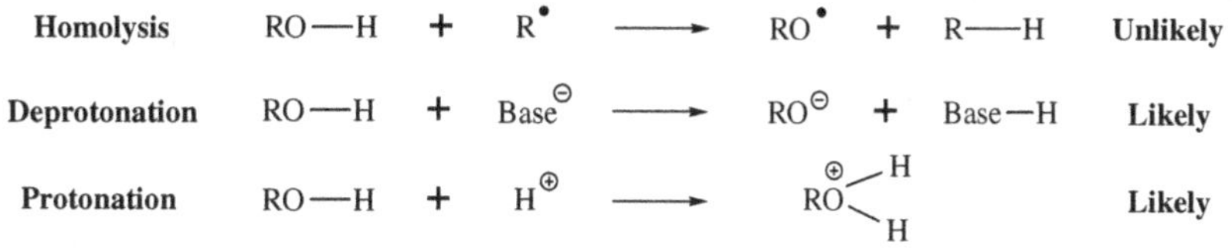

Homolysis	RO—H	+	R•	⟶	RO•	+	R—H	Unlikely
Deprotonation	RO—H	+	Base⊖	⟶	RO⊖	+	Base—H	Likely
Protonation	RO—H	+	H⊕	⟶	RO⊕—H / H			Likely

Fig. 5.3

As the reactivity of radicals is related to the bond strength, the best leaving groups in atom abstraction reactions are those that form a weak bond to carbon. Bromine and iodine are among the best leaving groups because the very weak carbon–halogen bond is readily cleaved to generate a carbon-centred radical (Fig. 5.4). This can be compared to nucleophilic substitution and elimination reactions for which bromide and iodide ions are also good leaving groups as both halogens can readily accommodate the negative charge. However, it is not always the case that good ionic leaving groups will also be good radical leaving groups. For example, the mesylate anion ($CH_3SO_3^-$) is a good leaving group in nucleophilic substitution and elimination reactions, but it cannot be used as a leaving group in radical reactions because the carbon–oxygen bond (R—OSO_2CH_3) is too strong.

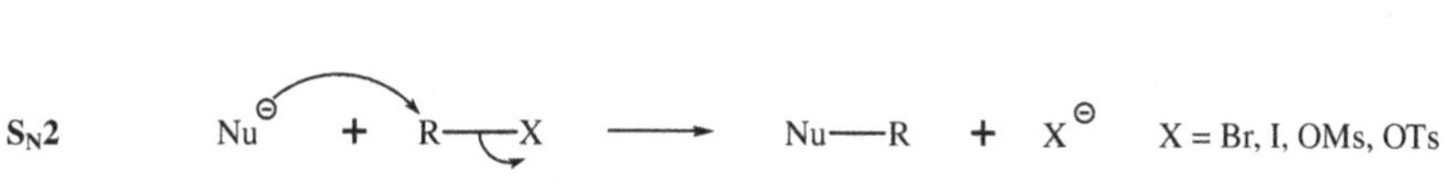

S_H2	R—X	+	R^1•	⟶	R•	+	X—R^1	X = Br, I
S_N2	Nu⊖	+	R—X	⟶	Nu—R	+	X⊖	X = Br, I, OMs, OTs

Ms = mesyl or methanesulfonyl; Ts = tosyl or 4-toluenesulfonyl

Fig. 5.4

5.2.4 Regioselectivity

The different reactivity of ions and radicals can lead to attack at different sites within a functional group. A classic example is the addition of hydrogen bromide to propene, shown in Fig. 5.5. Under polar (ionic) conditions, Markownikoff (Markovnikov) addition yields 2-bromopropane, whereas addition of peroxides (to promote radical formation) leads to the alternative regioisomer, 1-bromopropane, in an anti-Markownikoff addition (see Section 4.3.1). The different regioselectivities can be explained by the stability of the carbocation or radical intermediates; the proton adds to the alkene to give a more stable secondary (rather than primary) carbocation, while addition of the bromine radical proceeds to give the more stable secondary (rather than primary) radical.

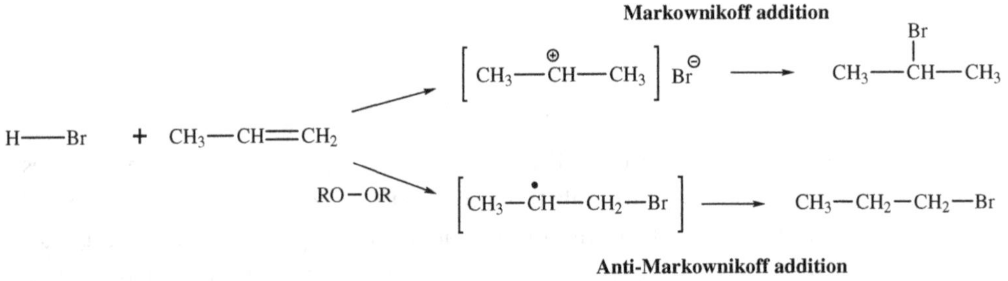

Markownikoff addition

$$H—Br \quad + \quad CH_3—CH{=\!=}CH_2$$

$$\left[CH_3—\overset{\oplus}{C}H—CH_3 \right] \overset{\ominus}{Br} \longrightarrow CH_3—\underset{\underset{Br}{|}}{C}H—CH_3$$

$$RO–OR \quad \left[CH_3—\overset{\bullet}{C}H—CH_2—Br \right] \longrightarrow CH_3—CH_2—CH_2—Br$$

Anti-Markownikoff addition

Fig. 5.5

Anionic reactions can also produce alternative regioisomers to those isolated from radical reactions. Organolithium and Grignard reagents generally add to α,β-unsaturated carbonyls at the carbonyl carbon (to give 1,2-addition), whereas carbon radicals, like organocopper reagents, attack the alkene carbon in a 1,4- (or Michael-type) addition (Fig. 5.6). Radicals add at the 4-position because the lowest unoccupied molecular orbital (LUMO) of the α,β-unsaturated carbonyl has the largest coefficient on the alkene carbon, and the larger the LUMO coefficient, the more readily an electron is accepted, i.e. the more likely a radical will add. We can also think of radicals as being 'soft' species because, like organocuprate reagents (R_2CuLi), which are soft nucleophiles (as they have the negative charge spread over a large transition metal), they prefer to attack the soft electrophilic centre at the alkene. In contrast, hard nucleophiles (including RLi and RMgX) have the negative charge localized on the carbon and these favour attack at the hard electrophilic centre, adjacent to the electronegative oxygen atom.

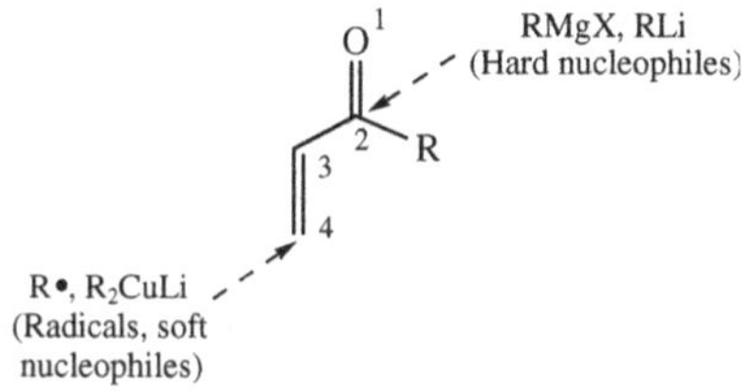

Fig. 5.6

Different regioselectivities are also observed in cyclization reactions and, whereas the hex-5-en-1-yl radical cyclizes to give the cyclopentane ring, the corresponding cation affords the cyclohexane ring (Fig. 5.7). The radical cyclization is under kinetic control and proceeds via a transition state in which the radical is closer to the internal alkene carbon (see Section 4.7.4). In contrast, the cation cyclization proceeds via a transition state in which the cation attacks the (electron-rich) centre of the double bond to produce the thermodynamically more stable secondary carbocation.

hex-5-en-1-yl radical — 5-*exo* product   hex-5-en-1-yl cation — 6-*endo* product

Fig. 5.7

5.2.5 Stereoselectivity

Reactions of acyclic (non-ring) cations or anions are generally more stereoselective than their radical counterparts. Bromine, for example, adds stereoselectively to (*E*)-2-butene because of an intermediate cyclic bromonium ion (Fig. 5.8). This ensures that the two bromine atoms add to the opposite sides of the alkene in an *anti* addition to give the (symmetrical) *meso*-dibromide.

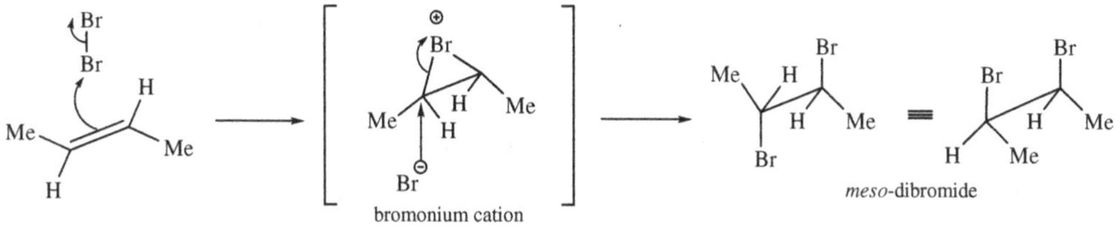

bromonium cation

meso-dibromide

Fig. 5.8

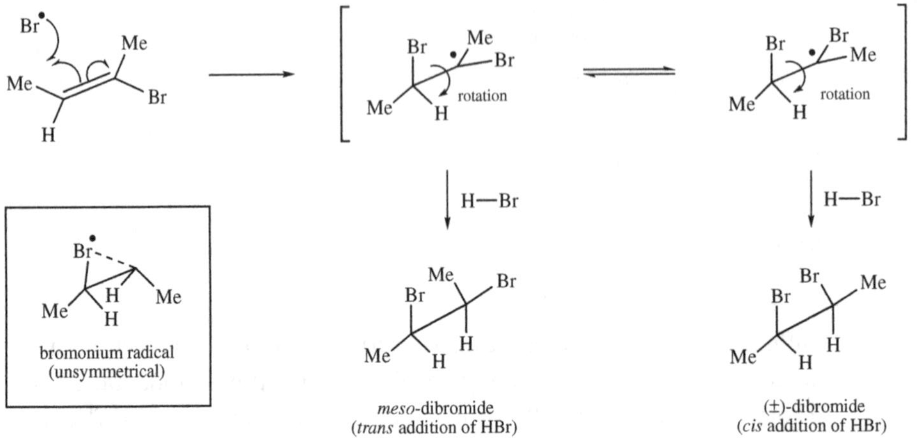

Fig. 5.9

When hydrogen bromide is added to a related alkene, (*E*)-2-bromobut-2-ene, in the presence of peroxides at room temperature, both *meso*- and unsymmetrical (±)-dibromides are formed (Fig. 5.9). This is because the central carbon–carbon bond of the intermediate tertiary radical can rotate (cf. the 'fixed' conformation of the bromonium ion). Subsequent hydrogen atom abstraction from the opposite face to the bulky bromine substituent leads to equal amounts of overall *cis* (same side) and *trans* (opposite side) addition of HBr to the alkene. At −80°C, however, only the *meso*-dibromide is observed and so the stereoselectivity improves as the temperature is lowered. This has been explained by the formation of a (unsymmetrical) bridged bromonium radical which, like the cyclic bromonium cation, restricts carbon–carbon bond rotation. In general, much better stereoselectivities are observed on reaction of cyclic rather than acyclic radicals, as these more rigid systems constrain the rotation of bonds leading to stereoselective reactions, particularly at low temperatures (see Section 7.4).

5.2.6 Polarity

Radicals and ions also show different substituent effects. For ionic chemistry, electron-donating substituents will aid the formation of electrophilic cations, whilst electron-withdrawing substituents will aid the formation of nucleophilic anions. An alkoxy substituent (RO) will therefore stabilize a cation, whereas a carbonyl group (RC=O) will stabilize an anion (Fig. 5.10).

For radicals, both electron-donating and electron-withdrawing substituents can stabilize the radical. In addition, the substituents will influence the polarity of the radical, and radicals adjacent to electron-donating substituents are nucleophilic, while radicals adjacent to electron-withdrawing substituents are electrophilic.

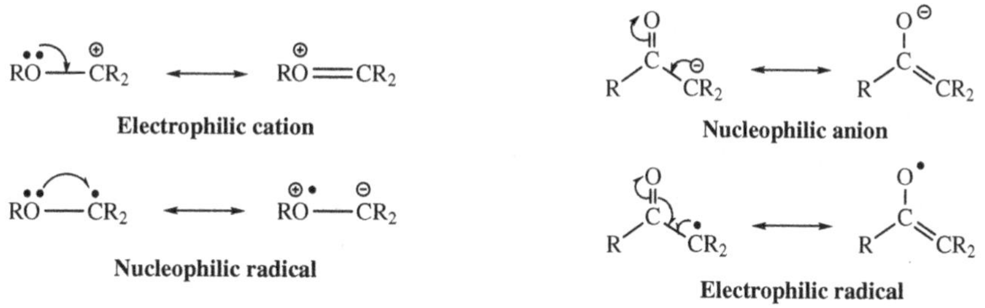

Fig. 5.10

The effect of the substituents leads to a reversal of reactivity (or *umpolung*). Whereas cations with alkoxy substituents prefer to add to electron-rich double bonds, radicals with alkoxy substituents (and a high-energy singly occupied molecular orbital, SOMO) are nucleophilic and so will prefer to add to (the LUMO of) electron-poor double bonds (Fig. 5.11).

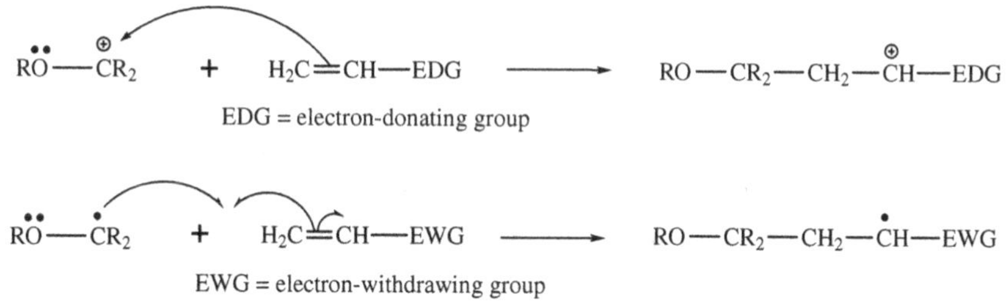

Fig. 5.11

5.3 Chain reactions

Most synthetically useful radical reactions involve chain processes. Following initiation, the radical reacts with a non-radical to produce a new radical in the propagation step (Fig. 5.12). As one radical is formed at the expense of another, a chain reaction occurs and this can be triggered by using only catalytic amounts of radical initiators. Typically 0.1 equivalent of the initiator to 1 equivalent of the radical precursor (RX) is used. Although only small (catalytic) amounts of the initiator are required to get the chain going, it is *not* a catalyst as it is destroyed during the reaction. Ideally, only one molecule of the initiator is required for complete conversion of R—X to product (R^2—X). However, competitive termination reactions (coupling or disproportionation) can destroy the intermediate radicals (R$^\bullet$, $R^{1\bullet}$ and $R^{2\bullet}$) and stop the chain. A steady supply of initiating radicals is required and this can be achieved by slow addition of the initiator to a solution of R—X.

The reaction temperature is important as this influences the rate of decomposition or half-life ($t_{1/2}$) of the initiator. The higher the temperature, the faster the rate of decomposition (and the lower the $t_{1/2}$), and reaction temperatures are usually adjusted so that $t_{1/2}$ is of the order of a few hours. This ensures that radicals are generated over a 'reasonable' time-scale.

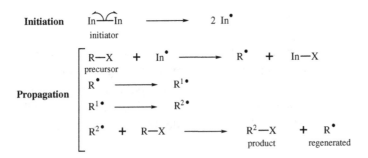

Fig. 5.12

Even though we can adjust the experimental conditions to ensure a low concentration of radicals (typically 10^{-7} mol dm^{-3}), a successful chain reaction still requires fast propagation steps, otherwise radical–radical termination steps will still predominate. We also need to have an understanding of the relative rates of the propagation steps because we require the radical intermediates to react selectively. For the conversion of R—X to R^2—X (Fig. 5.12), the $R^{1\bullet}$ radical must react to give $R^{2\bullet}$, otherwise reaction with R—X would lead to (unwanted) R^1—X. For $R^{1\bullet}$ and $R^{2\bullet}$ to react differently, they should have different substituents. If $R^{1\bullet}$ has an electron-donating substituent(s), then $R^{2\bullet}$ should ideally have an electron-withdrawing substituent(s)—this will ensure that the two radicals have opposite polarity and therefore very different reactivity.

Some reagents, often used to mediate radical reactions, will now be highlighted so as to illustrate the importance of chain reactions in synthesis.

5.3.1 Tributyltin hydride

n-Tributyltin hydride (Bu$_3$SnH) is the most popular reagent for conducting radical chain reactions. Rates of initiation and propagation reactions are well known, and many synthetically useful chain reactions to make carbon–carbon bonds have been developed using this reagent. Intra- or intermolecular reaction of alkyl or aryl halide, selenide (X = SePh) or sulfide (X = SPh) precursors to form reduced products (R^1—H) is possible as shown in Fig. 5.13.

Intramolecular   R—X + Bu$_3$SnH $\xrightarrow{\text{AIBN}\atop\text{0.1 equivalent}}$ R^1—H + Bu$_3$SnX

1 equivalent   1 equivalent added over > 1 h

Intermolecular   R—X + Bu$_3$SnH + Alkene $\xrightarrow{\text{AIBN}\atop\text{0.1 equivalent}}$ R^1—H + Bu$_3$SnX

1 equivalent   1 equivalent added over > 1 h   >10 equivalents

Fig. 5.13

Undesirable radical–radical termination steps are minimized by maintaining low concentrations of radicals, and good yields of product (R^1—H) can be isolated when using the 'standard' reagent ratios and addition times shown. These experimental conditions have developed from an understanding of the chain reaction mechanism and the need to minimize competing propagation reactions leading to by-products (Fig. 5.14).

Mechanism

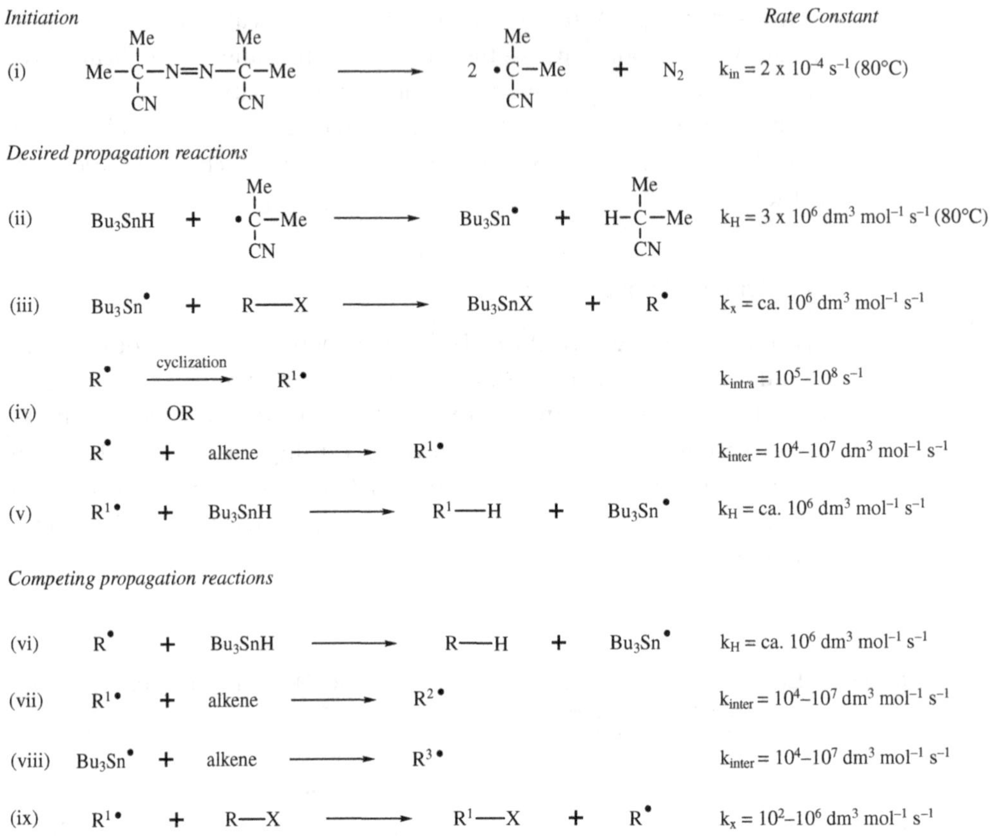

Initiation   *Rate Constant*

(i)
Me–C(Me)(CN)–N=N–C(Me)(CN)–Me $\longrightarrow$ 2 •C(Me)(CN)–Me + N$_2$ $k_{in} = 2 \times 10^{-4}$ s^{-1} (80°C)

Desired propagation reactions

(ii) Bu$_3$SnH + •C(Me)(CN)–Me $\longrightarrow$ Bu$_3$Sn• + H–C(Me)(CN)–Me $k_H = 3 \times 10^6$ dm^3 mol^{-1} s^{-1} (80°C)

(iii) Bu$_3$Sn• + R—X $\longrightarrow$ Bu$_3$SnX + R• $k_x =$ ca. 10^6 dm^3 mol^{-1} s^{-1}

(iv)
R• $\xrightarrow{\text{cyclization}}$ R^1• $k_{intra} = 10^5$–10^8 s^{-1}
OR
R• + alkene $\longrightarrow$ R^1• $k_{inter} = 10^4$–10^7 dm^3 mol^{-1} s^{-1}

(v) R^1• + Bu$_3$SnH $\longrightarrow$ R^1—H + Bu$_3$Sn• $k_H =$ ca. 10^6 dm^3 mol^{-1} s^{-1}

Competing propagation reactions

(vi) R• + Bu$_3$SnH $\longrightarrow$ R—H + Bu$_3$Sn• $k_H =$ ca. 10^6 dm^3 mol^{-1} s^{-1}

(vii) R^1• + alkene $\longrightarrow$ R^2• $k_{inter} = 10^4$–10^7 dm^3 mol^{-1} s^{-1}

(viii) Bu$_3$Sn• + alkene $\longrightarrow$ R^3• $k_{inter} = 10^4$–10^7 dm^3 mol^{-1} s^{-1}

(ix) R^1• + R—X $\longrightarrow$ R^1—X + R• $k_x = 10^2$–10^6 dm^3 mol^{-1} s^{-1}

Fig. 5.14

The initiation step (i) requires an azo compound, usually azobisisobuty-ronitrile (AIBN), which, on heating above 60°C or on photolysis, decomposes with the evolution of nitrogen to the isobutyronitrile radical. The rate of this initiation step is much lower than the rate of the subsequent propagation reactions and $t_{1/2}$ is approximately 1 h at 80°C. Even though this radical is tertiary and resonance stabilized, it is able to abstract rapidly a hydrogen atom from the tin hydride, a stronger C—H bond (360 kJ mol^{-1}) being formed at the expense of a weaker Sn—H bond (310 kJ mol^{-1}) [step (ii)]. As the isobutyronitrile radical is stabilized and not highly reactive, hydrogen atom abstraction from Bu$_3$SnH is selective and not observed for organic compounds which have stronger C—H bonds.

The tin radical, Bu$_3$Sn•, can then react with alkyl halides (except fluorides) to break the weak carbon–halogen bond ($\approx$ 230–350 kJ mol^{-1}) and form a stronger tin–halogen bond (289–393 kJ mol^{-1}) [step (iii)]. As might be expected, the rate of this abstraction reaction can vary considerably, and the weaker the carbon–halogen bond, the faster the reaction; tertiary iodides can react 10^7 times faster than primary chlorides. The stronger C—X bond in aryl and vinyl halides also means that chlorides are rarely used.

The resulting carbon-centred radical, $R^\bullet$, can then react through a series of inter- or intramolecular reactions to form a new radical $R^{1\bullet}$ [step (iv)]. The lifetimes of these radicals are determined by the rate of hydrogen atom abstraction (or reduction) from Bu_3SnH. To form the desired product R^1—H, radical $R^\bullet$ must react to form $R^{1\bullet}$ [step (v)], rather than react with Bu_3SnH to form R—H (a process known as simple reduction) in step (vi). For intramolecular radical reactions, this is less of a problem because unimolecular reactions are usually faster than bimolecular reactions with Bu_3SnH. We can also lower the concentration of tin hydride, by adding it slowly to the reaction mixture, as this will lower the rate of hydrogen atom trapping. Radical $R^\bullet$ will therefore have a longer lifetime, allowing more time for cyclization prior to trapping with tin hydride.

Cyclization	$R^\bullet \rightarrow R^{1\bullet}$	Rate $= k_{intra}[R^\bullet]$
		Favoured by low $[Bu_3SnH]$
Simple reduction	$R^\bullet + Bu_3SnH \rightarrow$	Rate $= k_H[R^\bullet][Bu_3SnH]$
	R—H $+ Bu_3Sn^\bullet$	

For the intermolecular reaction, we require $R^\bullet$ to react with, for example, an alkene before reaction with tin hydride. Simple reduction can be minimized by slow addition of tin hydride, and we can increase the rate of addition of $R^{1\bullet}$ to the alkene by using an excess of the alkene—this is usually present in 10–100-fold excess.

Intermolecular addition	$R^\bullet +$ alkene $\rightarrow R^{1\bullet}$	Rate $= k_{inter}[R^\bullet][$alkene$]$
		Favoured by high [alkene]
Simple reduction	$R^\bullet + Bu_3SnH \rightarrow$	
	R—H $+ Bu_3Sn^\bullet$	

If the concentration of the alkene is increased any further, $R^{1\bullet}$ will be more likely to react with another molecule of alkene (rather than with Bu_3SnH) to give $R^{2\bullet}$, and so on, to form a polymer [step (vii)]. This is not a problem if $R^\bullet$ and $R^{1\bullet}$ have different polarities, as their rates of reaction (with the same alkene) will be significantly different. A nucleophilic $R^\bullet$ radical, for example, will react with an electron-poor alkene much faster than would an electrophilic $R^{1\bullet}$ radical. The slow reaction of $R^{1\bullet}$ with an electron-poor alkene will favour reaction with tin hydride to give the desired product R^1—H, the rate of hydrogen atom transfer from Bu_3SnH

to alkyl radicals being much the same for nucleophilic and electrophilic radicals.

| Polymerization | $R^{1\bullet}$ + alkene → $R^{2\bullet}$, etc. | |
| Hydrogen atom transfer | $R^{1\bullet}$ + Bu_3Sn—H → R^1—H + $Bu_3Sn^{\bullet}$ | Favoured by a change in radical polarity |

Usually, tin radicals will prefer to abstract halogen atoms rather than add to alkene double bonds as this gives a stronger tin–halogen bond (289–393 kJ mol^{-1}) compared to the tin–carbon bond (270 kJ mol^{-1}). When an excess of the alkene is used, however, reaction of the alkene with the tin radical, $Bu_3Sn^{\bullet}$ (a process known as stannylation), can be problematic [step (viii)]. To minimize this, we need to use alkyl halide (R—X) precursors, such as iodides, with very weak bonds.

| Halogen atom transfer | $Bu_3Sn^{\bullet}$ + R—X → Bu_3Sn—X + $R^{\bullet}$ | Favoured by a weak C—X bond |
| Stannylation | $Bu_3Sn^{\bullet}$ + Alkene → $R^{3\bullet}$ | |

Finally, competing halogen atom abstraction by the carbon-centred radicals, $R^{\bullet}$ and $R^{1\bullet}$, should also be considered. Reaction of R—X with $R^{\bullet}$ is not a problem as this will regenerate the same starting materials, but reaction with $R^{1\bullet}$ will lead to a new halide R^1—X [step (ix)]. However, this is not a problem because any R^1—X that is formed will react with $Bu_3Sn^{\bullet}$ to regenerate $R^{1\bullet}$, and so no organohalide products are isolated from tin hydride reactions.

Tin hydride reactions with fast propagation steps and optimum reaction conditions can be very efficient and have long chain lengths. The propagation reactions may be repeated many thousands of times to give the desired product before competing termination reactions can take place. The ability to form carbon–carbon bonds in excellent yield, coupled with the very mild and selective nature of the reagent, has contributed to the popularity of this method in synthesis.

There are, however, limitations and one drawback is the need for a stoichiometric amount of tin hydride. If the tin hydride could be regenerated, then the method would produce smaller amounts of toxic tin halide by-products; this is particularly important, as separation of the tin halides from organic products can be difficult. One solution to this problem involves the use of sodium borohydride ($NaBH_4$) or the milder reducing agent sodium cyanoborohydride [$Na(CN)BH_3$]. These will reduce a tin–

halogen bond to a tin–hydrogen bond, and therefore a catalytic amount (0.1 equivalent) of the tin halide, together with 1 equivalent[1] of the hydride reducing agent, can be used. This ensures that the Bu_3SnH is present in low concentration, as only 0.1 equivalent of Bu_3SnH is formed initially and the rate of regeneration of Bu_3SnH (from Bu_3SnX) depends on the rate of formation of RX (Fig. 5.15).

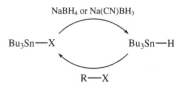

Fig. 5.15

5.3.2 Mercury hydrides

Mercury hydrides (RHgH), like tin hydrides, can be used to generate alkyl radicals. These hydrides are prepared *in situ* from reduction of alkylmercury halides or acetates (ethanoates) using $NaBH_4$ (Fig. 5.16).

$$R-HgX \xrightarrow{NaBH_4} R-HgH \qquad X = Br, Cl, OAc$$

Fig. 5.16

The Hg—H bond is extremely weak (≈ 60 kJ mol^{-1}), and therefore reaction with virtually any radical is expected to generate an alkylmercury(I) species which, on elimination of mercury metal, produces a carbon radical (Fig. 5.17). This can react with an alkene to form a new radical, and abstraction of a hydrogen atom from another molecule of RHgH completes the chain reaction. As the Hg—H bond is so weak, the rate of hydrogen atom abstraction is at least 10^7 dm^3 mol^{-1} s^{-1} at room temperature (which is around 10 times faster than for Bu_3SnH). To form R^1—H, the nucleophilic alkyl radical ($R^{\bullet}$) must therefore add rapidly to the alkene and this usually requires the presence of electron-withdrawing groups (e.g. CN, CO_2Et) on the double bond.

---

[1] 0.25 mole equivalent of $NaBH_4$ can be used because each molecule contains four hydrogen atoms.

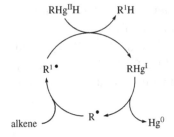

Fig. 5.17

The mild reaction conditions and easy access to a variety of organomercury precursors by, for example, oxymercuration of alkenes or transmetallation of Grignard reagents (Fig. 5.18) has led to many synthetic applications. However, the main drawback to this method is the toxicity of organomercury compounds and the disposal of mercury metal at the end of the reaction.

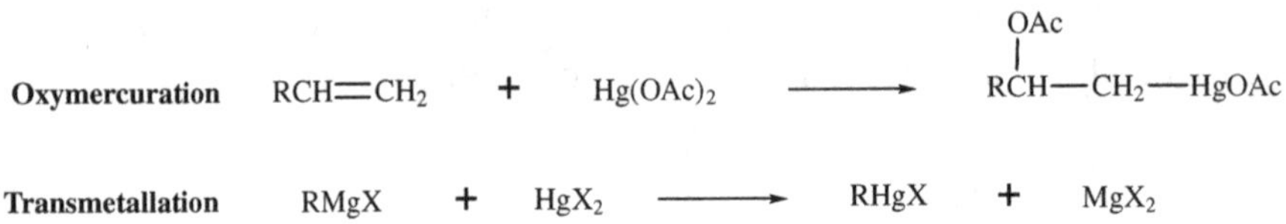

$$\textbf{Oxymercuration} \quad RCH\!=\!CH_2 \quad + \quad Hg(OAc)_2 \quad \longrightarrow \quad \overset{\displaystyle OAc}{\overset{|}{RCH}}\!-\!CH_2\!-\!HgOAc$$

$$\textbf{Transmetallation} \quad RMgX \quad + \quad HgX_2 \quad \longrightarrow \quad RHgX \quad + \quad MgX_2$$

Fig. 5.18

5.3.3 Silicon hydrides

The toxicity and purification problems associated with tin and mercury hydrides have led to the development of related hydrides, including those based on silicon. Silyl radicals, $R_3Si^\bullet$, react with organohalides, -selenides or -sulfides to generate alkyl radicals at similar rates to those observed for $Bu_3Sn^\bullet$ radicals (Fig. 5.19); the chain reaction sequence is analogous to that shown earlier for tin hydride (Fig. 5.14). However, the silicon–hydrogen bond in triorganosilanes (R_3Si—H) is stronger than that in tributyltin hydride by approximately 65 kJ mol^{-1}, and this means that reaction with alkyl radicals is slower, leading to short chain lengths at temperatures below 120°C (Fig. 5.19).

Halogen-atom abstraction

$$R_3Si^\bullet \quad + \quad X{-}R \quad \longrightarrow \quad R_3Si{-}X \quad + \quad {}^\bullet R \qquad k_x = \text{ca. } 10^6 \text{ dm}^3 \text{ mol}^{-1} \text{ s}^{-1}$$

$$Bu_3Sn^\bullet \quad + \quad X{-}R \quad \longrightarrow \quad Bu_3Sn{-}X \quad + \quad {}^\bullet R \qquad k_x = \text{ca. } 10^6 \text{ dm}^3 \text{ mol}^{-1} \text{ s}^{-1}$$

Hydrogen-atom abstraction

$$R^\bullet \quad + \quad H{-}SiR_3 \quad \longrightarrow \quad R{-}H \quad + \quad {}^\bullet SiR_3 \qquad k_H = \text{ca. } 10^3 \text{ dm}^3 \text{ mol}^{-1} \text{ s}^{-1}$$

$$R^\bullet \quad + \quad H{-}SnBu_3 \quad \longrightarrow \quad R{-}H \quad + \quad {}^\bullet SnBu_3 \qquad k_H = \text{ca. } 10^6 \text{ dm}^3 \text{ mol}^{-1} \text{ s}^{-1}$$

Fig. 5.19

   To overcome this problem, a catalytic amount of an alkyl thiol (RSH) can be added, as an alkyl radical will abstract the hydrogen atom of the thiol at a rate of approximately 10^6 dm^3 mol^{-1} s^{-1} at 25°C (Fig. 5.20). This leads to the formation of a carbon–hydrogen bond (≈ 410 kJ mol^{-1}) at the expense of a weaker sulfur–hydrogen bond (370 kJ mol^{-1}). As the Si—H bond is of similar strength (375 kJ mol^{-1}) to the S—H bond, the 1000-fold difference in the rates of hydrogen atom abstraction cannot be related to bond energies, but is best explained in terms of radical polarity. The sulfur reaction is fast because the nucleophilic alkyl radical rapidly abstracts the electron-poor hydrogen from the thiol; the hydrogen is electrophilic because it is bonded to the electronegative sulfur atom (RS$^{\delta-}$—H$^{\delta+}$). For the silicon reaction, the hydrogen is electron rich as it is bonded to an electropositive silicon atom, and hydrogen atom abstraction is consequently much slower (R$_3$Si$^{\delta+}$—H$^{\delta-}$). The resulting thiyl radical (RS$^\bullet$) is electrophilic as the electronegative sulfur prefers to accept an electron, and can abstract the electron-rich hydrogen atom from the silane. This reversed polarity facilitates the formation of a silyl radical even though the reaction is not exothermic (cf. similar energy of S—H and Si—H bonds). Although this is a reversible reaction (as polarity effects favour hydrogen atom transfer in either direction), the silyl radical can undergo competitive halogen atom abstraction, on reaction with R—X, to form a strong Si—X bond (320–470 kJ mol^{-1}).

$$R^\bullet \quad + \quad \overset{\delta+ \quad \delta-}{H{-}SR} \quad \longrightarrow \quad R{-}H \quad + \quad {}^\bullet SR$$
nucleophilic

$$RS^\bullet \quad + \quad \overset{\delta- \quad \delta+}{H{-}SiR_3} \quad \rightleftharpoons \quad RS{-}H \quad + \quad {}^\bullet SiR_3$$
electrophilic

Fig. 5.20

A slow single propagation step has therefore been replaced by two fast consecutive reactions, and this procedure is described as *polarity reversal catalysis*.

An alternative solution is to use a different silicon hydride with a weaker Si—H bond; this can be achieved by introducing bulky substituents on the silicon which, on homolysis, produces hindered and stable silyl radicals. This has lead to the development of tris(trimethylsilyl)silane (TTMSS), which is a less toxic and widely used alternative to tin hydride, despite being a more expensive reagent. The Si—H bond in this silane is only about 20 kJ mol^{-1} stronger than the Sn—H bond in tributyltin hydride, and this slightly stronger bond can be advantageous as it ensures a slower rate of hydrogen atom transfer (Fig. 5.21). Slow addition of the silane, to facilitate carbon–carbon bond formation, is therefore not so important because the lower rate of simple reduction means that the initial radical has more time to cyclize or add to an alkene. The silicon halide by-products are also more easily separated from the organic products and, as for tin hydride, catalytic amounts of TTMSS can be used if $NaBH_4$ is present to regenerate the reagent.

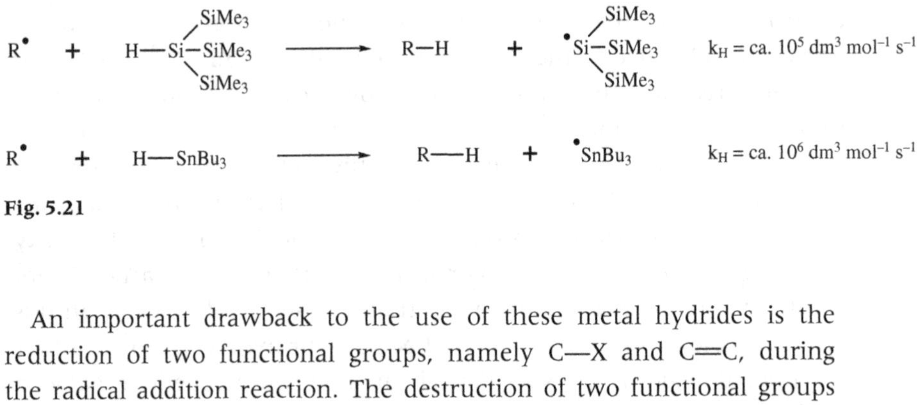

Fig. 5.21

An important drawback to the use of these metal hydrides is the reduction of two functional groups, namely C—X and C=C, during the radical addition reaction. The destruction of two functional groups in one reaction is wasteful and necessarily leads to products with reduced functionality; this, in turn, restricts the target molecules that can be prepared.

5.3.4 Halogen atom transfer

One strategy for generating more functionalized products involves halogen, rather than hydrogen, atom abstraction using alkyl halides. A catalytic amount of an initiator, such as hexabutylditin (Bu_6Sn_2), with a weak tin–tin bond (260 kJ mol^{-1}) can be used to give $Bu_3Sn^{\bullet}$ radicals which react with the alkyl halide to produce a small quantity of the precursor radical (R•) (Fig. 5.22). This radical can then undergo cyclization to give a product radical (R¹•), which is trapped by halogen atom transfer from another molecule of the alkyl halide. In the process, the precursor radical (R•) is regenerated and a product incorporating a versatile halogen atom (R¹X) is formed.

Initiation

$$2 \text{ R—X} \quad + \quad \text{Bu}_3\text{Sn—SnBu}_3 \quad \longrightarrow \quad 2 \text{ R}^\bullet \quad + \quad 2 \text{ Bu}_3\text{Sn—X}$$

Propagation

$$\text{R}^\bullet \xrightarrow{\text{cyclization}} \text{R}^{1\bullet}$$

$$\text{R}^{1\bullet} \quad + \quad \text{R—X} \quad \longrightarrow \quad \text{R}^\bullet \quad + \quad \text{R}^1\text{—X}$$

R• must be more stable than R¹•

Fig. 5.22

This reaction will only work if the carbon–carbon bond-forming reactions are fast, and therefore precursors with very weak C—X bonds must be used; the weaker the bond, the faster the halogen atom abstraction. For efficient halogen atom transfer, the product radical should be more reactive than the radical reactant and, in practice, this means that intramolecular cyclization reactions of iodides (with weak C—I bonds, $\approx 230 \text{ kJ mol}^{-1}$) work well. Cyclizations require that a strong σ bond is formed at the expense of a π bond, and these reactions are usually under kinetic control. Therefore, a secondary radical can cyclize to form a more reactive primary radical, which can rapidly abstract an iodine atom from the secondary iodide precursor leading to an efficient iodine atom chain transfer process.

Intermolecular halogen atom transfer can take place when polyhalogenated precursors, such as BrCCl_3, react with electron-rich alkenes in a Kharasch-type reaction (Fig. 5.23). The extremely weak C—Br (234 kJ mol^{-1}) bond in this compound can be cleaved on heating or photolysis to generate the electrophilic $^\bullet\text{CCl}_3$ radical (centred on a carbon with three electron-withdrawing chlorine substituents) and Br• radical. These add rapidly to the least hindered position of electron-rich alkenes to give carbon-centred radicals that preferentially abstract the bromine, rather than chlorine, atom from BrCCl_3. This is because of the weaker C—Br bond. Although Br• is generated in the initiation step, it is not formed in the subsequent propagation reactions and so the bromotrichloroalkane can be isolated in good yield.

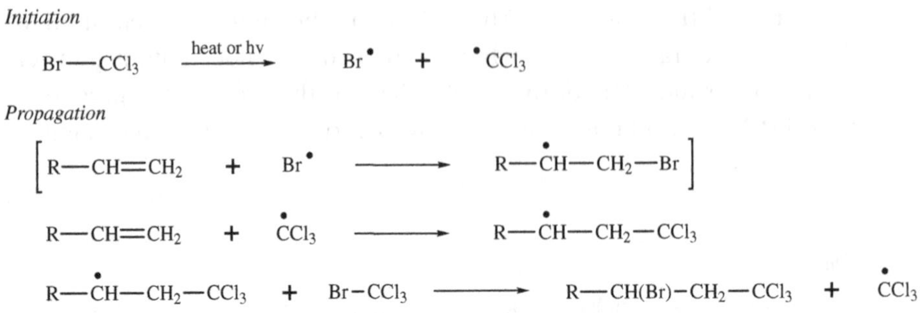

CHAPTER 5

Initiation

$$Br\!-\!CCl_3 \xrightarrow{\text{heat or hv}} Br^\bullet \;+\; {}^\bullet CCl_3$$

Propagation

$$\left[R\!-\!CH\!=\!CH_2 \;+\; Br^\bullet \longrightarrow R\!-\!\overset{\bullet}{C}H\!-\!CH_2\!-\!Br \right]$$

$$R\!-\!CH\!=\!CH_2 \;+\; {}^\bullet CCl_3 \longrightarrow R\!-\!\overset{\bullet}{C}H\!-\!CH_2\!-\!CCl_3$$

$$R\!-\!\overset{\bullet}{C}H\!-\!CH_2\!-\!CCl_3 \;+\; Br\!-\!CCl_3 \longrightarrow R\!-\!CH(Br)\!-\!CH_2\!-\!CCl_3 \;+\; {}^\bullet CCl_3$$

Fig. 5.23

Halogen atom transfer of polyhaloalkanes can also be promoted by the use of low-valent transition metal catalysts, such as copper(I) or ruthenium(II) complexes (Fig. 5.24). Inter- and intramolecular chain reactions are possible, and the success of the method relies on the easy interconversion of copper(I) and copper(II) and ruthenium(II) and ruthenium(III) oxidation states. Alkyl radicals are formed by halogen atom transfer to the metal (this may involve an electron transfer process) and these cyclize or add to alkenes to form more reactive radicals, $R^{1\bullet}$, which then abstract a halogen atom from the metal. As the metal is first oxidized and then reduced back to the same oxidation state, only catalytic quantities are required.

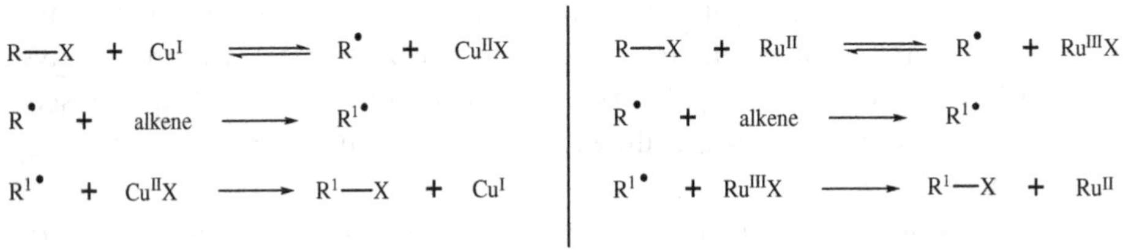

$$R\!-\!X \;+\; Cu^I \rightleftharpoons R^\bullet \;+\; Cu^{II}X \qquad\qquad R\!-\!X \;+\; Ru^{II} \rightleftharpoons R^\bullet \;+\; Ru^{III}X$$

$$R^\bullet \;+\; \text{alkene} \longrightarrow R^{1\bullet} \qquad\qquad R^\bullet \;+\; \text{alkene} \longrightarrow R^{1\bullet}$$

$$R^{1\bullet} \;+\; Cu^{II}X \longrightarrow R^1\!-\!X \;+\; Cu^I \qquad\qquad R^{1\bullet} \;+\; Ru^{III}X \longrightarrow R^1\!-\!X \;+\; Ru^{II}$$

Fig. 5.24

This method can be very efficient, and even intermolecular reactions produce little or no polymerization (derived from the addition of $R^{1\bullet}$ to the alkene) as the halogen atom transfer step to produce R^1X is very fast. This may be because the metal coordinates and so is in close proximity to the radical intermediates ($R^\bullet$ and $R^{1\bullet}$), thereby leading to a faster reaction with $Cu^{II}X$ or $Ru^{III}X$ than with a non-coordinated alkene molecule.

114

5.4 Non-chain reactions

The most synthetically useful non-chain reactions involve oxidative or reductive termination steps. In contrast to chain reactions, non-chain reactions require 1 equivalent of the radical initiator because the radical derived from the precursor (R•) is not regenerated during the reaction (Fig. 5.25). The propagation steps must be fast and products derived from radical–radical termination steps can be minimized by using low concentrations. These reactions also require selective oxidation or reduction of the product radical (R^{1}•); this can be achieved when R^{1}• has a different polarity, and therefore different substituents, to that of the radical reactant (R•).

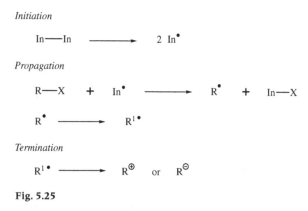

Initiation

$$In—In \longrightarrow 2\ In•$$

Propagation

$$R—X\ +\ In• \longrightarrow R•\ +\ In—X$$

$$R• \longrightarrow R^{1}•$$

Termination

$$R^{1}• \longrightarrow R^{\oplus}\ or\ R^{\ominus}$$

Fig. 5.25

5.4.1 Samarium(II) iodide—reduction

Samarium(II) iodide is a very powerful one-electron reducing agent and can initiate radical reactions by reduction of organohalides (Fig. 5.26). The formation of a carbon-centred radical is accompanied by oxidation of the metal to Sm^{III}, and this results in a blue (Sm^{II}) to yellow (Sm^{III}) colour change in tetrahydrofuran (THF) [step (i)]. Under these reducing conditions, the first-formed radical (R•) can be reduced by a further equivalent of samarium(II) iodide before the radical reaction (to give R^{1}•) can take place [step (ii)]. The desired radical reaction must therefore be quicker than reduction so that R^{1}• can be formed and subsequently reduced [step (iii)].

The reactions are usually carried out in THF, and a cosolvent, such as hexamethylphosphoramide (HMPA), is added to increase the reduction potential of the samarium(II); the more HMPA added, the faster the rate of reduction of the alkyl halide. This is because, in the absence of HMPA, only primary iodides and bromides are effectively reduced.

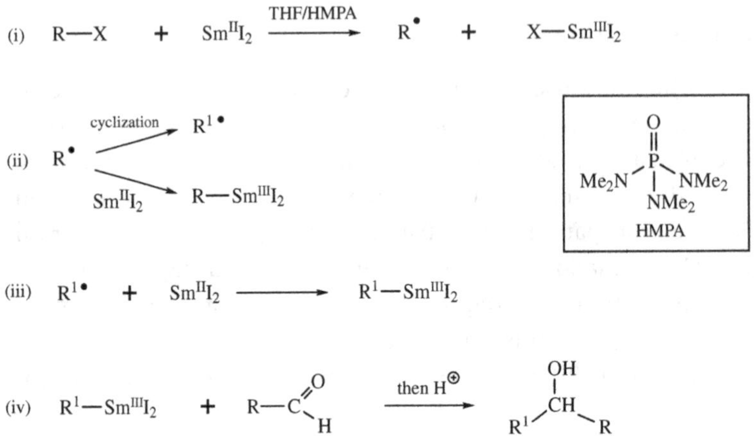

(i) $R-X$ + $Sm^{II}I_2$ $\xrightarrow{\text{THF/HMPA}}$ $R^{\bullet}$ + $X-Sm^{III}I_2$

(ii) $R^{\bullet}$ $\begin{array}{c} \xrightarrow{\text{cyclization}} R^{1\bullet} \\ \xrightarrow{Sm^{II}I_2} R-Sm^{III}I_2 \end{array}$

$$\begin{array}{c} O \\ \| \\ Me_2N-\overset{|}{\underset{|}{P}}-NMe_2 \\ NMe_2 \end{array}$$

HMPA

(iii) $R^{1\bullet}$ + $Sm^{II}I_2$ $\longrightarrow$ $R^{1}-Sm^{III}I_2$

(iv) $R^{1}-Sm^{III}I_2$ + $R-C\overset{O}{\underset{H}{\diagup}}$ $\xrightarrow{\text{then } H^{\oplus}}$ $R^{1}\diagup\overset{OH}{\underset{|}{CH}}\diagdown R$

Fig. 5.26

Although the use of HMPA extends the range of precursors which will react with samarium(II) iodide, it also increases the rate of reduction of the radical intermediates. When using 5–7 equivalents of HMPA to 1 equivalent of SmI_2, the rate of reduction of tertiary alkyl radicals is approximately 10^4 mol dm^{-3} s^{-1}, while the reduction of primary radicals, to give more stable primary carbanions, is faster ($\approx 10^6$ mol dm^{-3} s^{-1}). As these rates of reduction are very fast, this usually means that only intramolecular ($R^{\bullet}$ to $R^{1\bullet}$) radical cyclizations can be completed before reduction of $R^{\bullet}$. If we lower the concentration of HMPA, intermolecular radical reactions are still problematic because reduction of the product radical ($R^{1\bullet}$) is now so slow that radical polymerization and/or termination reactions can occur.

When samarium(II) reduces a radical, samarium(III) is formed which has a net positive charge; this may combine with the carbanion to form a covalent C—Sm bond [step (iii)]. Alternatively, the organosamarium ($R^{1}SmI_2$) product could be formed directly from the combination of $R^{1\bullet}$ with SmI_2. The organosamarium products can undergo traditional Grignard-type reactions, and therefore, for example, can react with aldehydes to give secondary alcohols after acidic work-up [step (iv)]. These 'tandem' sequences are very useful as two carbon–carbon bonds can be made in a 'one-pot' reaction. The first bond is formed by a radical cyclization reaction, while the second is derived from an ionic, nucleophilic addition reaction.

Samarium(II) iodide will also reduce aldehydes and ketones to give (ketyl) radical anions, which can be considered to be carbon-centred radicals because the samarium is oxyphilic and so likes to form a strong bond with oxygen [step (i), Fig. 5.27]. These radicals do not undergo further reduction in the presence of SmI_2 and therefore their lifetimes are significantly greater than those of alkyl radicals. Intermolecular additions as well as cyclizations are now possible, especially when electron-poor alkenes are used, because this ensures a fast reaction with the nucleophilic[2] ketyl radical [step (ii)]. The presence of an electron-withdrawing substituent on the alkene also means that addition leads to an electrophilic radical, and this is easily reduced using SmI_2 (often in the absence of cosolvents) [step (iii)].

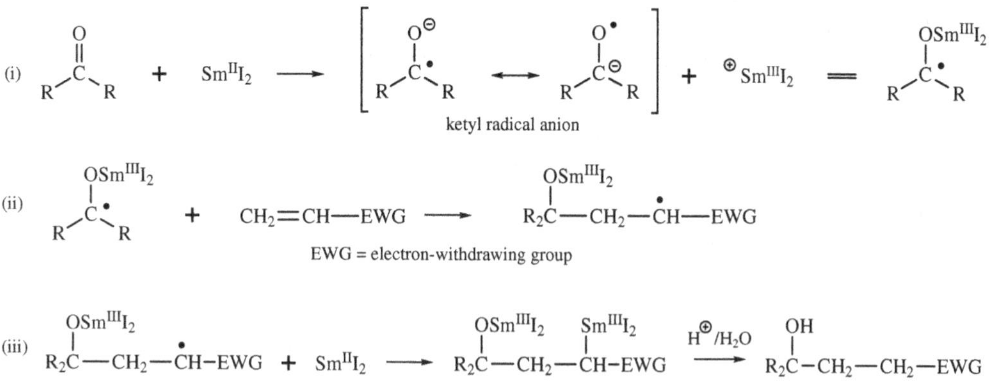

ketyl radical anion

EWG = electron-withdrawing group

Fig. 5.27

Particularly good yields of products can be isolated from sequential reactions involving an initial ketyl radical cyclization. The ability to carry out a variety of cyclizations and quench with a number of different electrophiles makes this a powerful and versatile synthetic method. Highly stereoselective cyclizations are also possible because of chelation using samarium(III); this can 'fix' the geometry of the radical precursor to give a rigid transition state for cyclization (see Section 7.4). There are some disadvantages as the reagent is very reactive and needs to be handled under inert atmospheres (to prevent oxidation) with extra safety precautions because natural samarium contains two radioactive isotopes (which are both weak α-emitters). In addition, reactions requiring HMPA are particularly hazardous as this is a carcinogenic compound and precautions must be taken to avoid exposure.

---

[2] The radical is nucleophilic because the unpaired electron can interact with a lone pair on oxygen.

5.4.2 Cobalt(I) complexes—reduction

Coenzyme B_{12}, which is the biochemically active form of vitamin B_{12} (cyanocobalamin), is known to orchestrate a variety of radical reactions in nature. These include a number of molecular rearrangements for which the mechanism involves an initial homolytic cleavage of the weak cobalt–carbon bond in the coenzyme. A number of related, but simpler, alkylcobalt(III) complexes (including cobaloximes, salens and salophens) have since been prepared, and these have also been shown to undergo radical reactions. These derivatives can be prepared, for example, by reaction of an alkyl halide with cobalt(I) in a nucleophilic substitution reaction [step (ii), Fig. 5.28]. The reactive cobalt(I) 'anion' is generated *in situ* from reduction of the more stable cobalt(II) or cobalt(III) oxidation states [step (i)], and the chloro(pyridine)cobaloxime(III) complex, for example, can be reduced either electrochemically at the cathode or by reaction with $NaBH_4$. Following homolysis of the cobalt(III)–carbon bond [step (iii)], the carbon-centred radical can react intra- or intermolecularly to form a new carbon-centred radical [step (iv)], which is then trapped by cobalt(II) [step (v)]. The product alkylcobalt(III) complex can then undergo a β-elimination (or dehydrocobaltation) reaction to form an alkene product together with a cobalt(III) hydride [step (vi)]. If the reaction is carried out in an electrochemical cell, for example, then the cobalt(III) hydride can be reduced back to cobalt(I) at the cathode, and so only a catalytic amount of the complex is required.

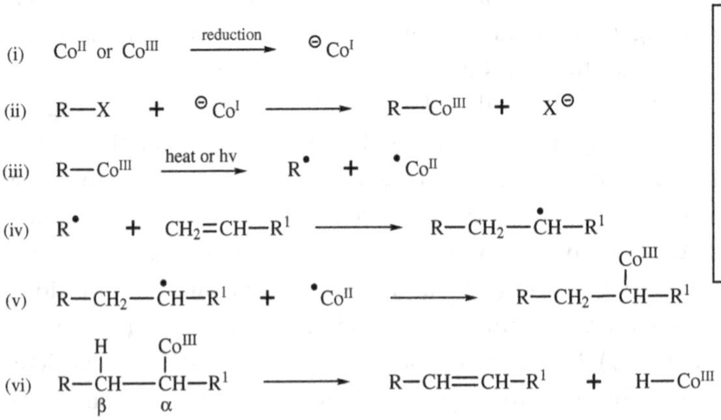

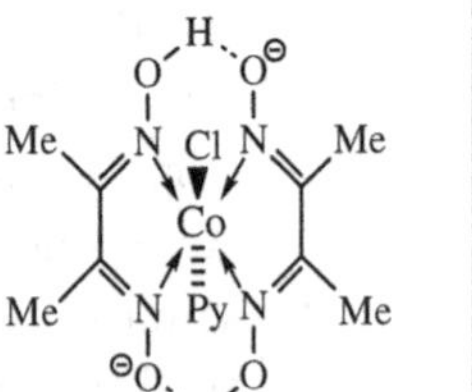

(i) Co^{II} or Co^{III} $\xrightarrow{\text{reduction}}$ $^\ominus Co^I$

(ii) $R—X$ + $^\ominus Co^I$ $\longrightarrow$ $R—Co^{III}$ + $X^\ominus$

(iii) $R—Co^{III}$ $\xrightarrow{\text{heat or hv}}$ $R^\bullet$ + $^\bullet Co^{II}$

(iv) $R^\bullet$ + $CH_2{=}CH—R^1$ $\longrightarrow$ $R—CH_2—\overset{\bullet}{C}H—R^1$

(v) $R—CH_2—\overset{\bullet}{C}H—R^1$ + $^\bullet Co^{II}$ $\longrightarrow$ $R—CH_2—\overset{Co^{III}}{\underset{|}{C}H}—R^1$

(vi) $R—\overset{H}{\underset{\beta}{\overset{|}{C}H}}—\overset{Co^{III}}{\underset{\alpha}{\overset{|}{C}H}}—R^1$ $\longrightarrow$ $R—CH{=}CH—R^1$ + $H—Co^{III}$

Chloro(pyridine)cobaloxime(III)

Fig. 5.28

These cobalt group transfer reactions are synthetically useful because the alkene product contains a versatile double bond. In some cases, the intermediate organocobalt(III) species can undergo alternative reactions to give other functionalized products. These include alcohols, which can be formed on reaction of RCo^{III} with molecular oxygen followed by (reduction using) NaBH$_4$.

The use of only a catalytic quantity of the cobalt complex is very attractive, but this requires the selective reduction of the cobalt(III) hydride (in the presence of an alkyl halide); this can restrict the type of alkyl halide precursors and reducing agents that can be used. A further drawback can be dehydrocobaltation of the initial alkylcobalt(III) species (RCo^{III}). If the radical addition step is slow, premature β-elimination can occur to form predominantly, or even exclusively, a 'simple' alkene (cf. simple reduction for Bu$_3$SnH).

5.4.3 Manganese(III) acetate—oxidation

Manganese(III) acetate is a one-electron oxidant that reacts with carbonyls, and particularly 1,3-dicarbonyls (which have an acidic α-hydrogen), to produce radicals (Fig. 5.29). The mechanism of radical generation is not completely understood, but is believed to involve the formation of an intermediate manganese(III) enolate [step (i)]. This is thought to react to give manganese(II), together with an electrophilic β-diketo radical (with two adjacent electron-withdrawing carbonyl groups), which is not easily oxidized by a further equivalent of manganese(III)—this would produce a very unstable carbocation. The electrophilic radical can therefore add intra- or intermolecularly to electron-rich double bonds and, the more electron rich the double bond, the faster the rate of addition [step (ii)]. This produces a nucleophilic radical that can be readily oxidized by a further equivalent of manganese(III) in an oxidative termination step [step (iii)]. The cation can subsequently undergo the loss of a proton to form an alkene or react with a nucleophile and, as reactions are generally conducted in acetic (ethanoic) acid, and the manganese has acetate ligands, acetate (ethanoate) products are often isolated.

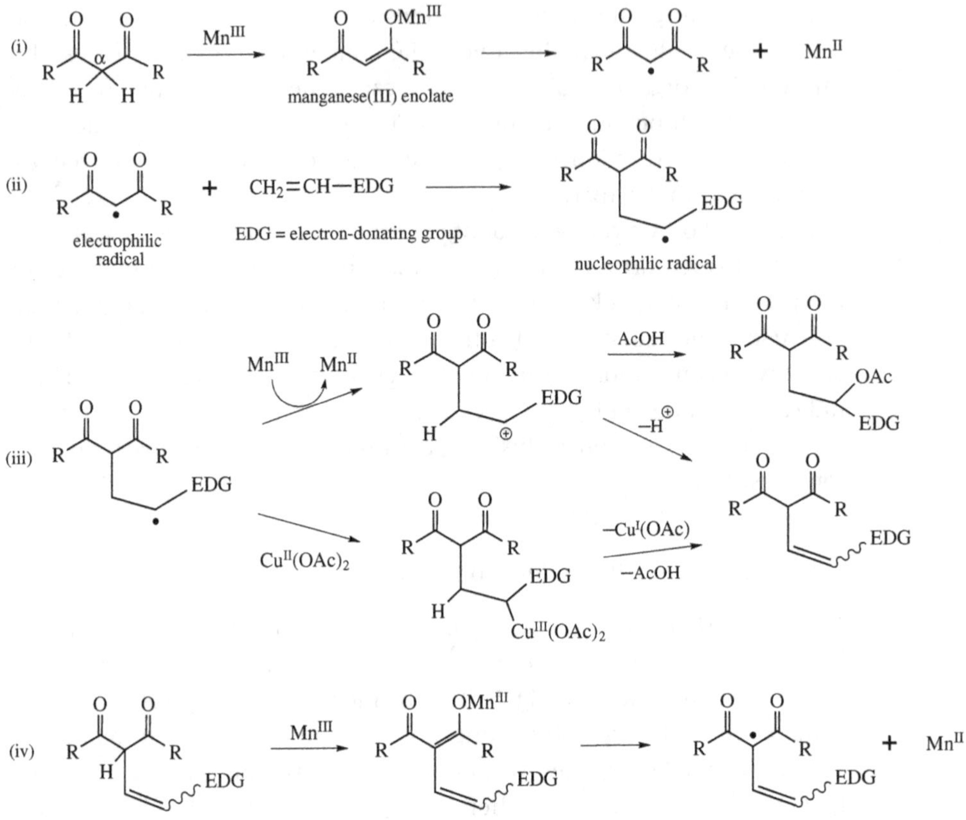

Fig. 5.29

Although Mn(OAc)₃ readily oxidizes tertiary alkyl radicals, a more powerful oxidizing agent, usually Cu(OAc)₂, is often added to efficiently oxidize primary and secondary radicals [step (iii)]. This ensures that the product radicals are oxidized before any radical termination reactions can take place. Intermediate organocopper(III) complexes may be formed, which can lead to an alkene on loss of copper(I) [cf. cobalt(III) complexes] and acetic acid.

The use of inexpensive manganese and copper acetates to produce functionalized products, such as alkenes, by oxidative termination is synthetically attractive. However, the method is limited to carbonyl precursors (which readily enolize) and the reactions involve at least 2 equivalents of the metal oxidant. This initially limited its use on an industrial scale, but an electrochemical oxidation has since been developed so as to regenerate the manganese(III) from manganese(II) *in situ*, and as little as 0.2 equivalent of manganese(III) can be used in some cases. One further concern is that products which contain α-hydrogen atoms can react further with manganese(III) to produce another manganese enolate [step (iv), Fig. 5.29]. This can lead to further radical generation and oxidative termination to give mixtures of products, particularly when the product is more easily oxidized than the starting material.

5.5 Summary

Radical reactions complement the traditional ionic methods used in organic synthesis. Indeed, there are a number of advantages to using radical intermediates: these include the mild, neutral reaction conditions which allow acid- or base-sensitive molecules to be transformed without decomposition. The rates of a variety of elementary radical reactions are now known and synthetic chemists can make use of these to design complex radical processes. Chain reactions are particularly elegant, requiring only a small amount of the initiator, and very efficient transformations (with long chain lengths) can be routinely carried out using hydrogen or halogen atom transfer reactions. Although non-chain processes require a stoichiometric amount of the initiator, they can be used to access functionalized products by a combination of radical chemistry followed by ionic chemistry in 'one-pot' reactions.

Further reading

Baguley, P.A. & Walton, J.C. (1998) Flight from the tyranny of tin: the quest for practical radical sources free from metal encumbrances. *Angewandte Chemie, International Edition in English*, **37**, 3072–3082.

Ballestri, M., Chatgilialoglu, C., Clark, K.B., Griller, D., Giese, B. & Kopping, B. (1991) Tris(trimethylsilyl)silane as a radical-based reducing agent in synthesis. *Journal of Organic Chemistry*, **56**, 678–683.

Curran, D.P. (1988a) The design and application of free radical chain reactions in organic synthesis. Part 1. *Synthesis*, 417–439.

Curran, D.P. (1988b) The design and application of free radical chain reactions in organic synthesis. Part 2. *Synthesis*, 489–513.

Jasperse, C.P., Curran, D.P. & Fevig, T.L. (1991) Radical reactions in natural product synthesis. *Chemical Reviews*, **91**, 1237–1286.

Melikyan, G.G. (1998) Manganese-based organic and bioinorganic transformations. *Aldrichimica Acta*, **31**, 50–64.

Molander, G.A. & Harris, C.R. (1996) Sequencing reactions with samarium(II) iodide. *Chemical Reviews*, **96**, 307–338.

Molander, G.A. & Harris, C.R. (1998) Sequencing reactions with samarium(II) iodide. *Tetrahedron*, **54**, 3321–3354.

Pattenden, G. (1988) Cobalt-mediated radical reactions in organic synthesis. *Chemical Society Reviews*, **17**, 361–382.

Ramaiah, M. (1987) Radical reactions in organic synthesis. *Tetrahedron*, **43**, 3541–3676.

Russell, G.A. (1989) Free radical chain reactions involving alkyl- and alkenylmercurials. *Accounts of Chemical Research*, **22**, 1–8.

Snider, B.B. (1996) Manganese(III)-based oxidative free-radical cyclizations. *Chemical Reviews*, **96**, 339–363.

Walling, C. (1985) Some properties of radical reactions important in synthesis. *Tetrahedron*, **41**, 3887–3900.

CHAPTER 6

Functional Group Transformations

6.1 Introduction

The synthesis of most target molecules normally involves a number of reactions that are used to introduce and/or transform functional groups. Often these transformations are used to prepare molecules which can subsequently react in carbon–carbon bond-forming reactions. However, this is not always easy, and the ability to introduce or transform only one functional group in a structurally complex molecule is a considerable synthetic challenge. For radical reactions, we need to selectively break the weakest bond in the precursor to give a radical that reacts via only one pathway. This can be difficult with reactive radicals, but it does offer the possibility of introducing a functional group in even very unreactive molecules. Alkanes, for example, are inert in ionic reactions, but they can be halogenated or oxidized in radical processes. This chapter discusses these important transformations and also highlights a number of alternative functional group interconversions that involve very selective radical reactions.

6.2 Transformations

6.2.1 Halogenation

The conversion of an alkane to an alkyl halide is a very useful transformation. Halogenations using molecular chlorine or, particularly, bromine are the most common because fluorinations are violent and difficult to control, while iodinations are not thermodynamically possible because the carbon–iodine bond is so weak.

Although the chlorination of alkanes using chlorine radicals shows some selectivity (Fig. 6.1), these reactions are usually hard to control and generally produce a mixture of products.

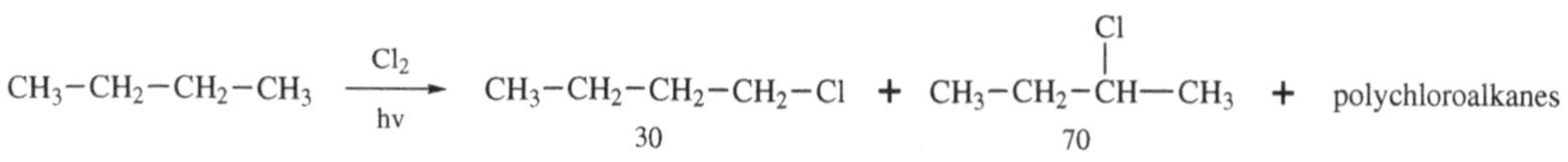

$$CH_3-CH_2-CH_2-CH_3 \xrightarrow[h\nu]{Cl_2} \underset{30}{CH_3-CH_2-CH_2-CH_2-Cl} \; + \; \underset{70}{CH_3-CH_2-\overset{\overset{\displaystyle Cl}{|}}{CH}-CH_3} \; + \; \text{polychloroalkanes}$$

Fig. 6.1

To overcome the problem of regioselectivity, a symmetrical alkane, such as cyclopentane, with only a single type of C—H bond can be used (Fig. 6.2). If the alkane is used in excess, polychlorination can also be minimized.

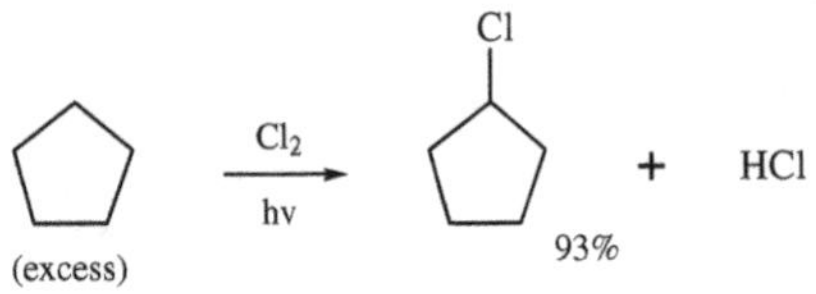

Fig. 6.2

Precursors containing a particularly weak C—H bond can also react selectively and, for example, α,α,α-trichlorotoluene (or benzotrichloride) can be readily prepared from toluene (Fig. 6.3). Chlorination occurs at the methyl side chain because the benzylic radical is more stable than an aryl radical, and so the methyl C—H bonds are much weaker than those on the aromatic ring. Even so, the reactive chlorine radicals can also add to the benzene ring and this has been used to prepare the insecticide Lindane or Gammexane (γ-BHC) (Fig. 6.4). A mixture of stereoisomers is produced from this reaction and the desired γ-isomer is only isolated in 13–18% yield.

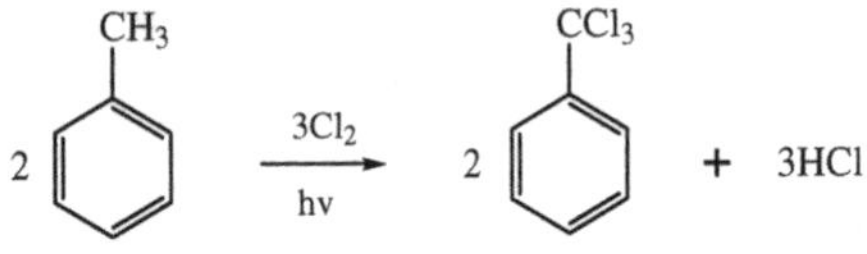

Fig. 6.3

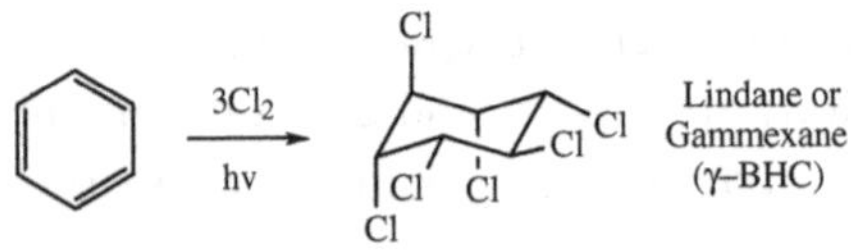

Fig. 6.4

Radical brominations are more selective than chlorinations (as Br• is less reactive than Cl•), and allylic (or benzylic) brominations are particularly selective. These reactions involve the abstraction of a hydrogen atom at the carbon adjacent to the double bond to produce a resonance-stabilized allylic radical (Fig. 6.5a). The allylic radical reacts with a molecule of bromine to give an allylic bromide together with a bromine radical. The bromine radical can then abstract a hydrogen atom from another molecule of alkene to continue the chain reaction. These types of reaction are known as substitutions because a bromine atom in the product substitutes a hydrogen atom in the starting material.

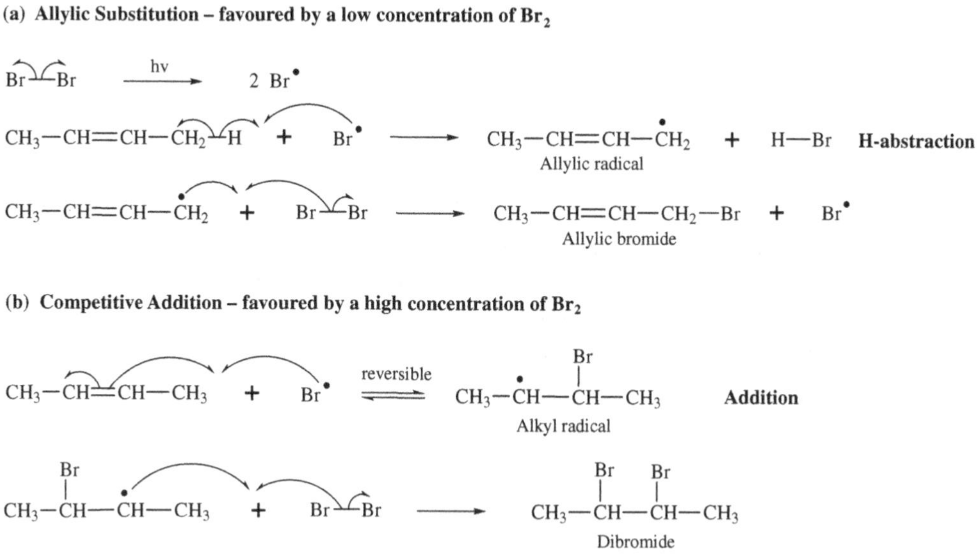

(a) Allylic Substitution – favoured by a low concentration of Br$_2$

$$Br \frown Br \xrightarrow{\;hv\;} 2\ Br^\bullet$$

$$CH_3-CH=CH-CH_2 \frown H \ + \ Br^\bullet \longrightarrow CH_3-CH=CH-\overset{\bullet}{C}H_2 \ + \ H-Br \quad \textbf{H-abstraction}$$
Allylic radical

$$CH_3-CH=CH-\overset{\bullet}{C}H_2 \ + \ Br \frown Br \longrightarrow CH_3-CH=CH-CH_2-Br \ + \ Br^\bullet$$
Allylic bromide

(b) Competitive Addition – favoured by a high concentration of Br$_2$

$$CH_3-CH=CH-CH_3 \ + \ Br^\bullet \underset{}{\overset{reversible}{\rightleftharpoons}} \ CH_3-\overset{\bullet}{C}H-\overset{\overset{\displaystyle Br}{|}}{C}H-CH_3 \quad \textbf{Addition}$$
Alkyl radical

$$CH_3-\overset{\overset{\displaystyle Br}{|}}{C}H-\overset{\bullet}{C}H-CH_3 \ + \ Br \frown Br \longrightarrow CH_3-\overset{\overset{\displaystyle Br}{|}}{C}H-\overset{\overset{\displaystyle Br}{|}}{C}H-CH_3$$
Dibromide

Fig. 6.5

Unfortunately, addition reactions can compete with allylic substitutions. Thus, bromine can add to alkenes in an ionic reaction via a bromonium ion and this can be minimized by using non-polar solvents (e.g. CCl$_4$); however, bromine radicals can also add to alkenes to form dibromides (Fig. 6.5b). This competitive reaction is more important for allylic rather than benzylic bromination, because an isolated double bond is more reactive than the conjugated double bonds in benzene. However, good yields of allyl bromides can be isolated when using a low concentration of Br$_2$. This is because the bromine radical adds *reversibly* to a carbon–carbon double bond. Therefore, the formation of the dibromide requires a high concentration of Br$_2$ to ensure that the intermediate alkyl radical reacts with Br$_2$ before the reverse reaction (β-elimination) can take place. In contrast, however, the hydrogen atom abstraction reaction to give an allylic radical is irreversible. This means that the allylic radical has a much longer lifetime than the alkyl radical, and therefore if Br$_2$ is added slowly to the reaction mixture it will selectively react with the allylic radical.

Although substitutions can be carried out by slow addition of Br$_2$, the reactions are generally performed using the less hazardous

N-bromosuccinimide (NBS). This reagent effectively generates a low concentration of Br_2 by an ionic reaction with hydrogen bromide (Fig. 6.6).

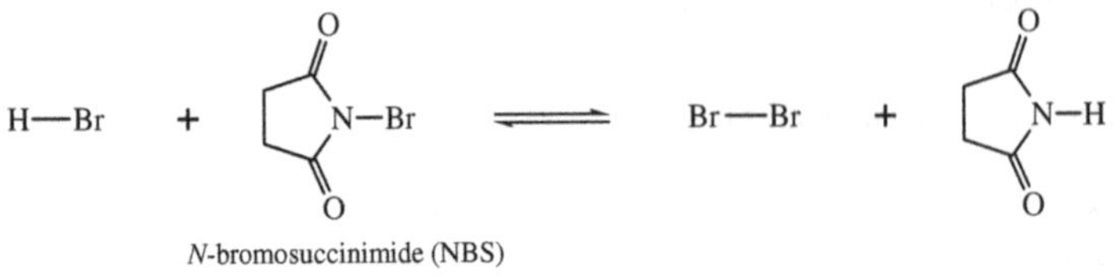

N-bromosuccinimide (NBS)

Fig. 6.6

There is usually a trace of Br_2 in the NBS that initiates the reaction to form the allylic radical and hydrogen bromide. The hydrogen bromide then reacts with NBS to regenerate the Br_2 required for reaction with the allylic radical. Therefore, every time one molecule of hydrogen bromide is formed, one molecule of Br_2 is subsequently produced, and a low concentration of Br_2 is therefore maintained throughout the reaction.

6.2.2 Dehalogenation

There are a variety of methods available for reducing a carbon–halogen to a carbon–hydrogen bond. A number of ionic reactions are known using metal hydrides, including sodium borohydride ($NaBH_4$) or the more reactive lithium aluminium hydride ($LiAlH_4$). These nucleophilic substitution reactions involve attack by the hydride ion (H^-), which is bound to the metal, and primary halides react by an S_N2 mechanism (substitution, nucleophilic, bimolecular) (Fig. 6.7a). Primary halides often give better yields than secondary or tertiary halides, which are less reactive and more susceptible to elimination to give alkene products.

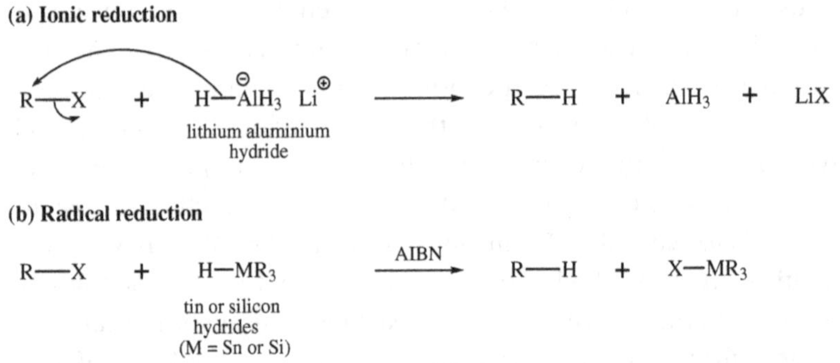

(a) Ionic reduction

lithium aluminium
hydride

(b) Radical reduction

tin or silicon
hydrides
(M = Sn or Si)

Fig. 6.7

The most common radical methods employ tin or silicon hydrides together with a radical initiator (usually azobisisobutyronitrile, AIBN) (Fig. 6.7b). A chain reaction develops in which the metal-centred radical abstracts the halogen atom in an S_H2 reaction (substitution, homolytic, bimolecular) (see Section 5.3.1). Primary, secondary and tertiary precursors can be used with iodine, bromine or chlorine leaving groups and, in contrast to ionic hydride reductions, the radical reactions work best with tertiary halides as these have the weakest carbon–halogen bonds.

The radical reactions are generally much more selective than the ionic metal hydride reactions. Chemoselective reduction of the carbon–halogen bond can be carried out in the presence of many other functional groups (including carbonyls) that would also be reduced using, for example, $LiAlH_4$. The neutral reaction conditions also prevent the formation of alkenes, which are often formed under the basic conditions of $NaBH_4$ and $LiAlH_4$ reductions.

6.2.3 Oxygenation

Many organic compounds can be selectively oxidized by molecular oxygen (a diradical) at temperatures below 100°C in a process known as autoxidation. At higher temperatures, the reactions are hard to control and complete oxidation of all carbons, to give carbon dioxide, can take over in a process known as combustion. In organic synthesis, the most useful reaction of this type involves the selective oxidation of a compound containing a weak carbon–hydrogen bond to give a hydroperoxide (Fig. 6.8). Although hydroperoxides are very reactive compounds and not usually isolated, they can be used as intermediates in synthetically useful processes.

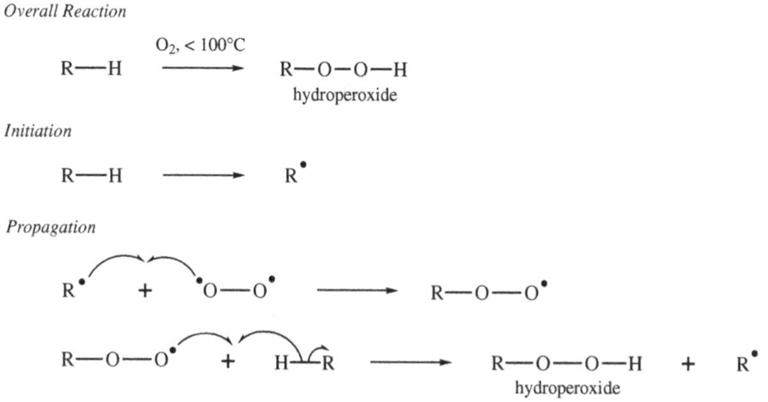

Overall Reaction

$$R\text{—}H \xrightarrow{\text{O}_2, < 100°C} R\text{—}O\text{—}O\text{—}H$$
hydroperoxide

Initiation

$$R\text{—}H \longrightarrow R^{\bullet}$$

Propagation

$$R^{\bullet} + {}^{\bullet}O\text{—}O^{\bullet} \longrightarrow R\text{—}O\text{—}O^{\bullet}$$

$$R\text{—}O\text{—}O^{\bullet} + H\text{—}R \longrightarrow R\text{—}O\text{—}O\text{—}H + R^{\bullet}$$
hydroperoxide

Fig. 6.8

127

The reaction can be represented by a chain reaction mechanism which is initiated by homolytic cleavage of the C—H bond. Although peroxides increase the rate of oxidation, their presence is not essential, and the compound can be oxidized simply by heating with molecular oxygen. As molecular oxygen is not able to abstract a hydrogen atom from the substrate, the formation of an alkyl radical, in the initiation step, is not well understood and is often attributed to trace metal impurities. However, once formed, the carbon-centred radical reacts extremely rapidly with oxygen to give a peroxyl radical (ROO•), which is able to selectively abstract a hydrogen atom from another molecule of substrate (to complete the chain reaction). The abstraction reactions are selective because the peroxyl radical is not particularly reactive (as the radical is stabilized by the lone pair on the adjacent oxygen atom) and so forms a relatively weak O—H bond in the hydroperoxide (368 kJ mol^{-1}; cf. 435 kJ mol^{-1} for an alcohol O—H bond). This essentially means that the reaction will generally take place at tertiary, allylic and benzylic positions as these have carbon–hydrogen bonds which are (slightly) weaker than 368 kJ mol^{-1}.

This process can be carried out on a large scale, and phenol is produced industrially from cumene (or 1-methylethylbenzene) via a hydroperoxide (Fig. 6.9). The hydroperoxide is selectively formed at the benzylic position, and this is then treated with dilute acid to give phenol (and propanone) by an acid-catalysed rearrangement. Tetralin can also be oxidized to form a similar hydroperoxide which, in the presence of base, undergoes elimination to form α-tetralone.

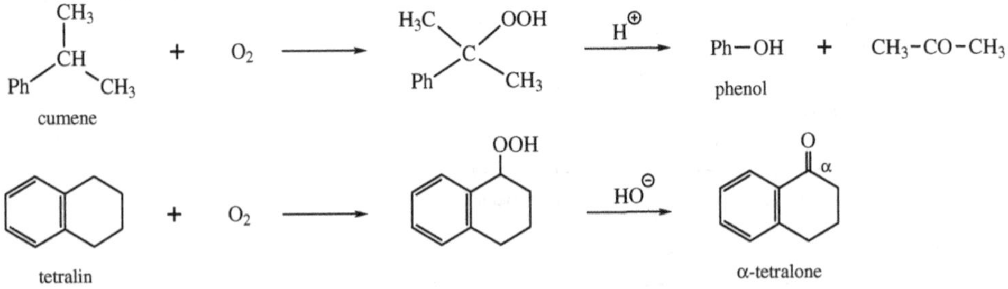

Fig. 6.9

It should be stressed that autoxidation can also be a nuisance. Ethereal solvents are susceptible to oxidation at the α-position to form hydroperoxides even at room temperature (Fig. 6.10). These are explosive when heated, and so tetrahydrofuran (THF) or diethyl ether should not be distilled before washing with a reducing agent (such as $FeSO_4$) to remove any accumulated hydroperoxides.

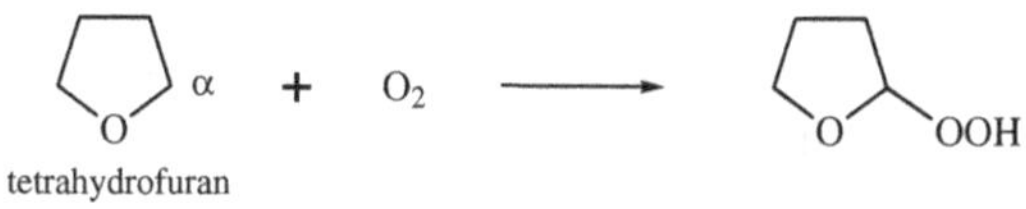

tetrahydrofuran

Fig. 6.10

Aldehydes are particularly susceptible to oxidation on standing in air at room temperature (Fig. 6.11). Benzaldehyde (PhCHO), for example, regularly needs to be purified by distillation to remove the white crystals of benzoic acid which gradually accumulate in samples exposed to air. This autoxidation reaction can be initiated by a catalytic amount of a metal oxidant; homolysis of the relatively weak aldehyde C—H bond (360 kJ mol^{-1}) generates an acyl radical. This acyl radical then reacts with oxygen to give an acylperoxyl radical ($RCO_3{}^{\bullet}$), which can abstract a hydrogen atom from another molecule of the aldehyde. This chain reaction produces a peracid, which is a very powerful oxidizing agent, and it is this that reacts with the aldehyde (in an ionic disproportionation reaction) to give the carboxylic acid product.

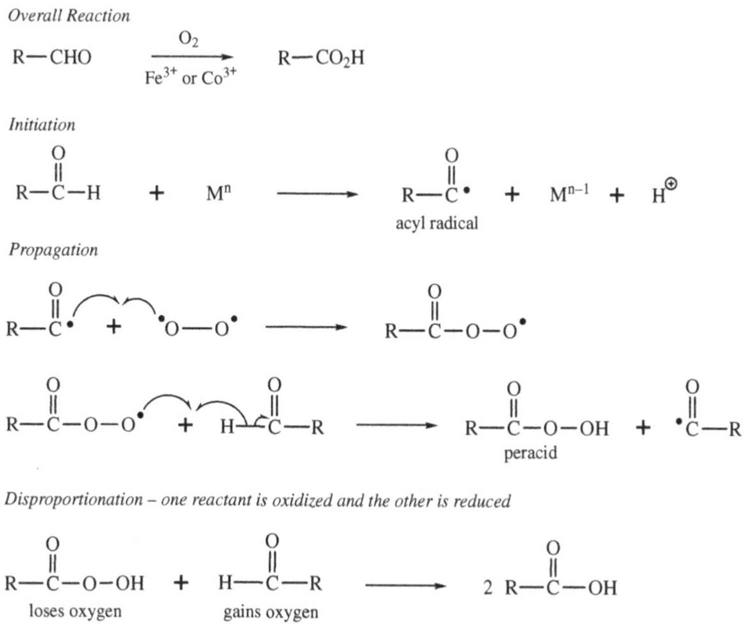

Overall Reaction

Initiation

Propagation

Disproportionation – one reactant is oxidized and the other is reduced

Fig. 6.11

129

6.2.4 Deoxygenation

The reduction of an alcohol to an alkane is usually difficult to achieve in one step. Apart from benzyl-type alcohols, which can undergo catalytic hydrogenolysis, alcohols need to be activated and the OH group converted to a better leaving group. For ionic transformations, this is often achieved by conversion to a tosylate which is readily reduced by $LiAlH_4$ or $NaBH_4$ (Fig. 6.12).

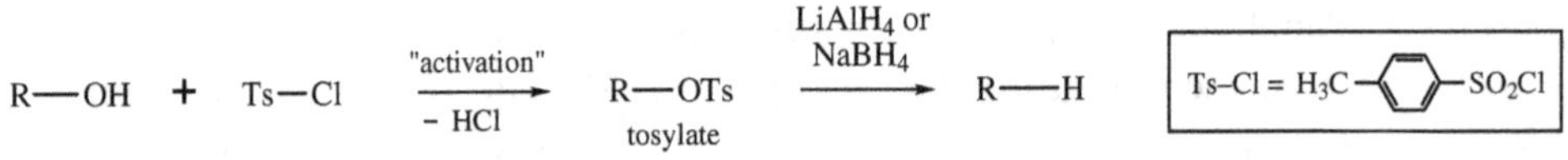

Fig. 6.12

A similar activation step is required for radical transformations because it is difficult to break the strong carbon–oxygen bond (R—OH). A common method of activation involves conversion to xanthates (or related thiocarbonyls with a C=S bond), and these can be prepared from alcohols using a three-step reaction sequence (Fig. 6.13).

Activation

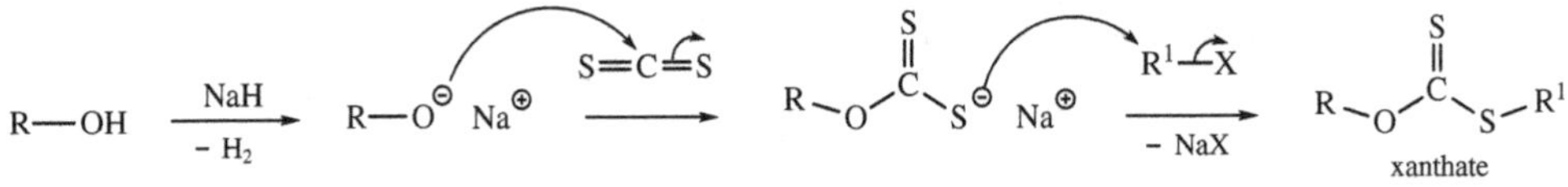

Fig. 6.13

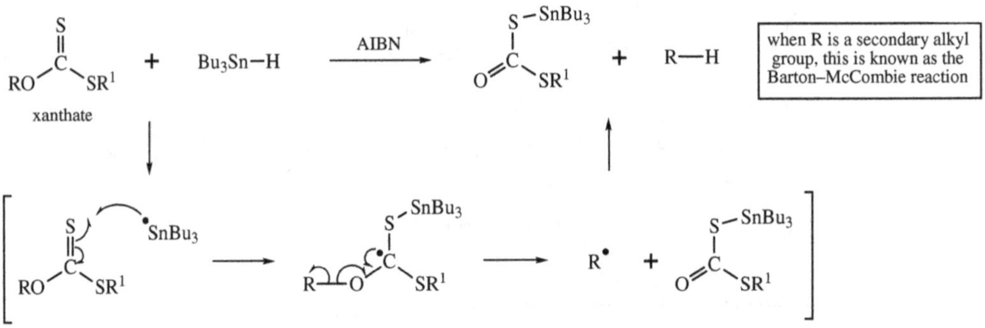

| when R is a secondary alkyl group, this is known as the Barton–McCombie reaction |

Fig. 6.14

When xanthates are reacted with tributyltin hydride and a radical initiator, the tributyltin radical adds to the carbon–sulfur double bond to give a tertiary radical that fragments (Fig. 6.14). The alkyl radical produced abstracts a hydrogen atom from the tin hydride to regenerate the tributyltin radical and so complete the chain reaction. During the reaction, the xanthate C=S bond is broken and replaced by a much stronger C=O bond in the tin by-product—this is the driving force for the reaction. The efficiency of the reaction depends on the rate of β-fragmentation of the tertiary radical. Xanthates derived from secondary alcohols (R = secondary alkyl group) react more quickly and under milder conditions than those derived from primary alcohols because the fragmentation produces a more stable radical.

This is a very chemoselective method of reduction, as shown by the reaction of xanthate (11), derived from a natural (taxane) product called baccatin III, to give the 7-deoxy analogue (12) in excellent yield (Fig. 6.15).

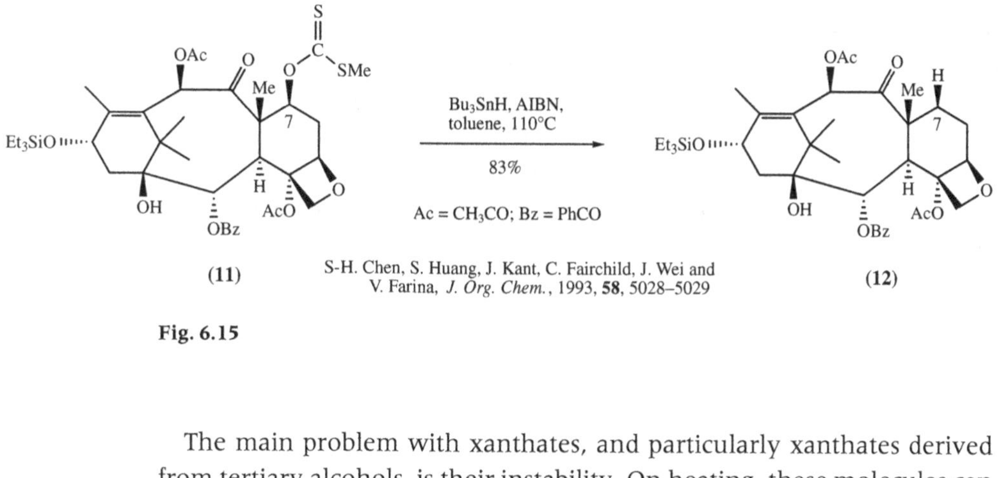

Bu_3SnH, AIBN,
toluene, 110°C

83%

Ac = CH_3CO; Bz = PhCO

(11)   (12)

S-H. Chen, S. Huang, J. Kant, C. Fairchild, J. Wei and
V. Farina, *J. Org. Chem.*, 1993, **58**, 5028–5029

Fig. 6.15

The main problem with xanthates, and particularly xanthates derived from tertiary alcohols, is their instability. On heating, these molecules can undergo an intramolecular elimination to form an alkene in the Chugaev reaction (Fig. 6.16).

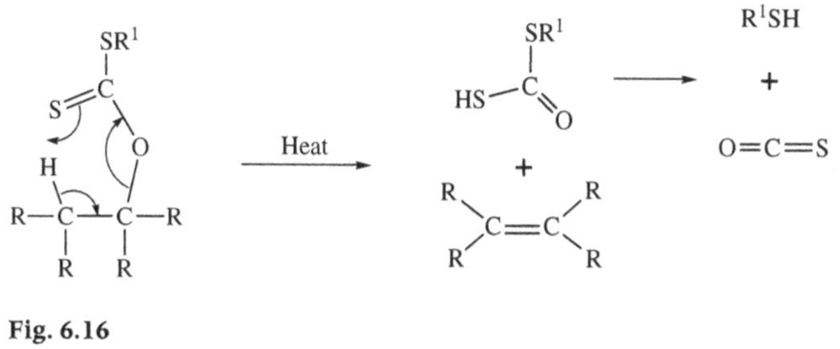

Heat

Fig. 6.16

A solution to this problem involves the use of thiohydroxamic ester derivatives, prepared from the reaction of alcohols with oxalyl chloride, followed by N-hydroxypyridin-2-thione (Fig. 6.17).

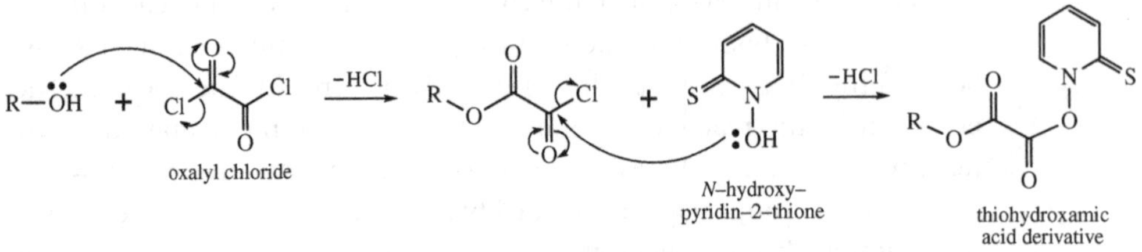

oxalyl chloride

N–hydroxy–
pyridin–2–thione

thiohydroxamic
acid derivative

Fig. 6.17

Reaction with the tributyltin radical produces a chain reaction which leads to the deoxygenated product together with a pyridine by-product (Fig. 6.18). This reaction is favoured by aromatization (as an aromatic pyridine ring is produced), the formation of two strong C=O bonds and entropy, as four products are generated from the two reactants. The reaction is particularly effective for the deoxygenation of tertiary alcohols (as the radical derived from tin radical addition to the thiohydroxamic acid derivative rapidly fragments to form a stable tertiary alkyl radical intermediate), and so this nicely complements the xanthate deoxygenation reaction.

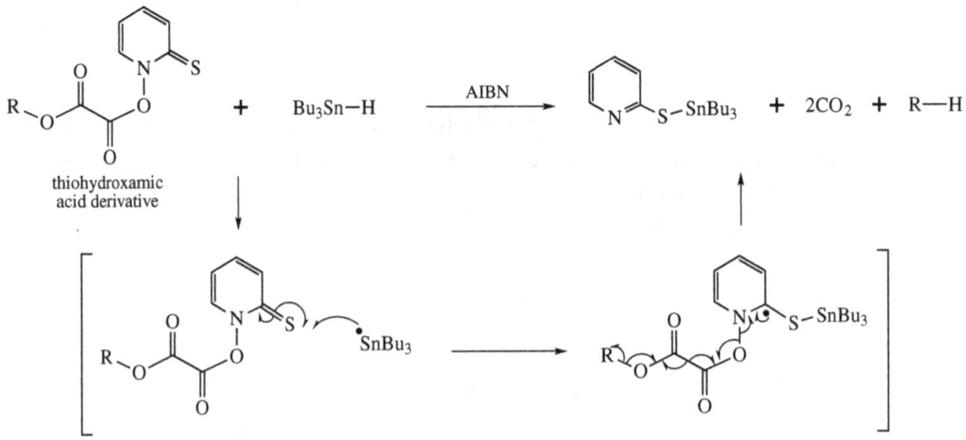

thiohydroxamic
acid derivative

Fig. 6.18

6.2.5 Decarboxylation

Carboxylic acids can lose carbon dioxide on treatment with base, or simply on heating, but neither of these are general methods (Fig. 6.19). The anionic reaction works best with electron-withdrawing substituents (which are able to stabilize the carbanion), while thermal decarboxylation is limited to acids, such as β-keto acids, which can adopt a cyclic, six-membered transition state. These limitations, however, can be overcome using radical methods, all of which involve an initial 'activation' of the carboxylic acid group.

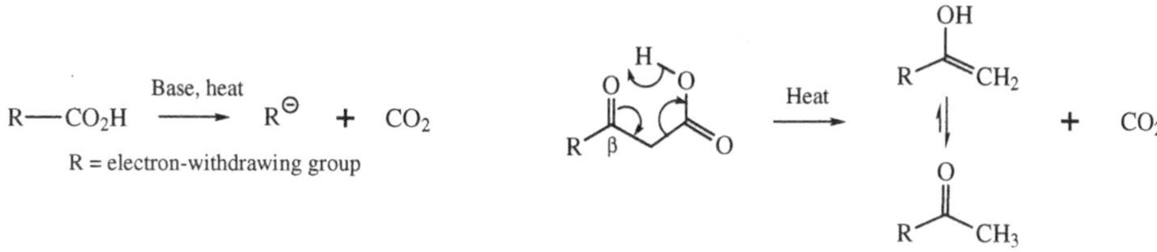

Fig. 6.19

Kolbe, who showed that carboxylic acid salts could be converted to alkanes in an electrochemical cell, developed the first method of radical decarboxylation. The carboxylic acid is activated by deprotonation and the carboxylate anion is then oxidized to the carboxyl radical, which then fragments to lose carbon dioxide (Fig. 6.20). Even reactive primary radicals (R$^{\bullet}$) can be generated, and these can combine to form alkanes (R—R) with an even number of carbon atoms on dimerization.

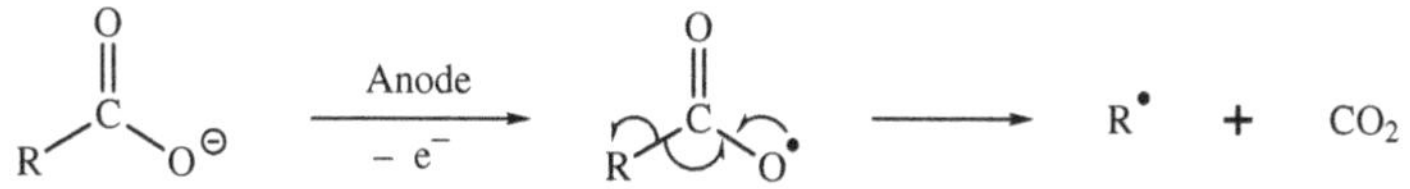

Fig. 6.20

Hunsdiecker subsequently developed a related method, starting from silver carboxylates. These are reacted with bromine to form an acyl hypobromite (in an ionic reaction) with a weak O—Br bond (Fig. 6.21). This undergoes homolysis to give the acyloxyl or carboxyl radical ($RCO_2^{\bullet}$), which fragments to give an alkyl radical and carbon dioxide. The alkyl radical then abstracts a bromine atom from a second molecule of acyl hypobromite to carry on the chain reaction. Although the reaction is of wide scope (as it can be used to prepare primary, secondary and tertiary bromides), this method has now been superseded by the use of thiohydroxamic esters in the Barton decarboxylation reaction. This avoids the use of expensive silver salts and hazardous liquid bromine.

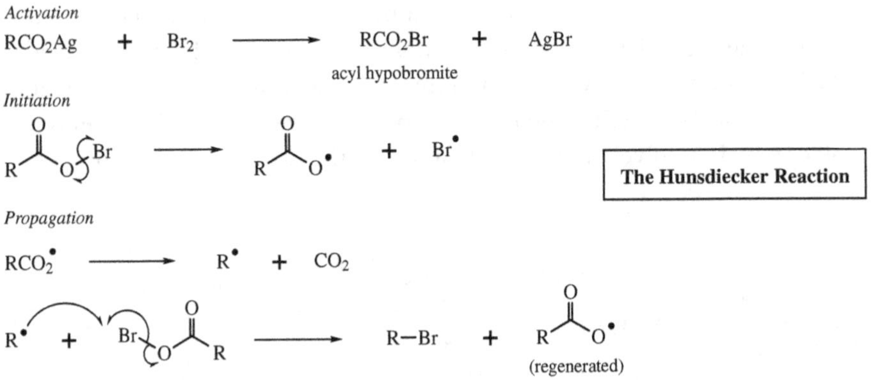

Activation

$$RCO_2Ag \quad + \quad Br_2 \quad \longrightarrow \quad RCO_2Br \quad + \quad AgBr$$

acyl hypobromite

Initiation

The Hunsdiecker Reaction

Propagation

$$RCO_2^{\bullet} \quad \longrightarrow \quad R^{\bullet} \quad + \quad CO_2$$

Fig. 6.21

Thiohydroxamic esters are readily prepared from reaction of *N*-hydroxypyridin-2-thione with acid chlorides derived from carboxylic acids (Fig. 6.22). These react with tributyltin hydride in a chain reaction, similar to that shown earlier for the deoxygenation of alcohols (see Section 6.2.4), to produce the reduced organic product together with a pyridine by-product and carbon dioxide. This is a general reaction, and primary, secondary and tertiary carboxylic acids can all be reduced via decarboxylation of the intermediate carboxyl radical.

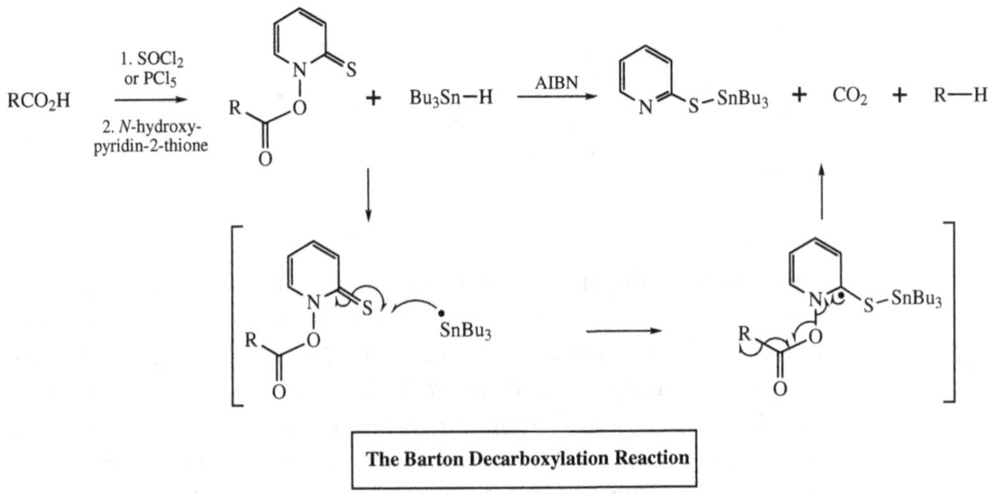

The Barton Decarboxylation Reaction

Fig. 6.22

Thiohydroxamic esters can also react with other radical initiators, and heating with bromotrichloromethane, for example, produces alkyl bromides (Fig. 6.23). This chain reaction works because the trichloromethyl radical ($^\bullet CCl_3$), like $Bu_3Sn^\bullet$, is 'thiophilic', and likes to form a bond with sulfur to give, after fragmentation, an alkyl radical which abstracts the bromine atom from $BrCCl_3$ (to regenerate $^\bullet CCl_3$).

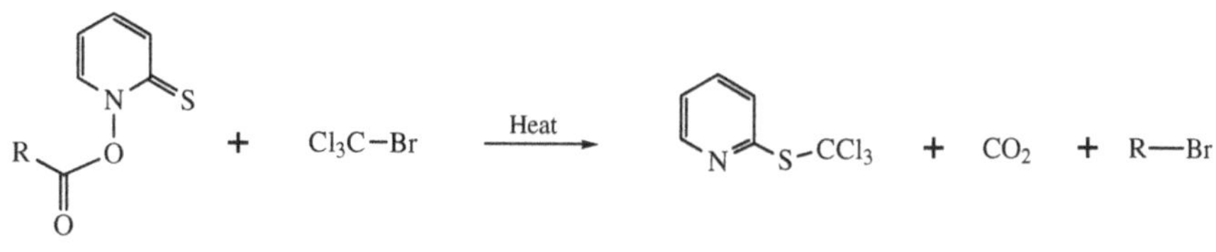

Fig. 6.23

6.2.6 Carbonylation and cyanation

The decarboxylation reactions highlighted in Section 6.2.5 are effective because acyloxyl radicals ($RCO_2^\bullet$) fragment to form alkyl radicals and carbon dioxide in a fast exothermic reaction. This means that the reverse reaction is not possible even when alkyl radicals are reacted with a high concentration of carbon dioxide. In contrast, decarbonylation reactions are less exothermic and are reversible; therefore, alkyl radicals can react to form acyl radicals under a high pressure of carbon monoxide (Fig. 6.24). This process can be used to form aldehydes from alkyl halides in a one-carbon homologation reaction. Alkyl radicals are formed from reaction between tributyltin hydride and primary, secondary or tertiary halides, and these can all react with carbon monoxide to give acyl radicals ($RCO^\bullet$). The acyl radicals abstract a hydrogen atom from the tributyltin hydride to regenerate the tributyltin radical and form the aldehyde product, generally in excellent yield.

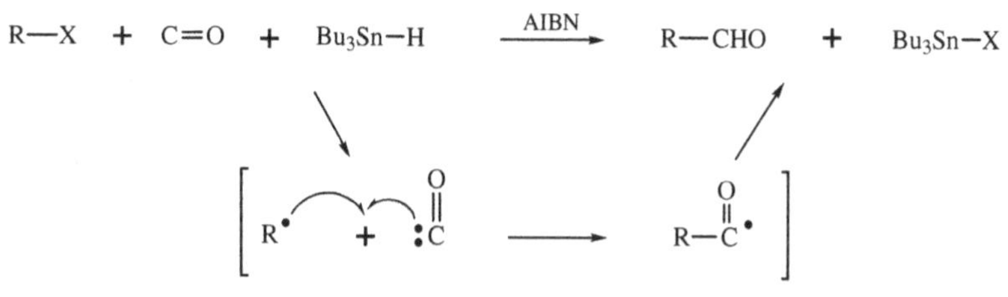

Fig. 6.24

An alternative and much more common one-carbon homologation reaction involves the conversion of alkyl halides to nitriles (Fig. 6.25). This is generally carried out using sodium or potassium cyanide; the cyanide ion displaces the halide in an S_N2 reaction to give good yields of nitriles using primary, allylic or benzylic halides.

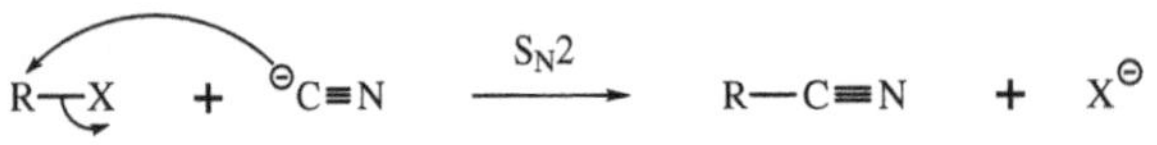

$$R\text{---}X \quad + \quad {}^{\ominus}C\equiv N \quad \xrightarrow{S_N2} \quad R\text{---}C\equiv N \quad + \quad X^{\ominus}$$

Fig. 6.25

$$R\text{---}X \quad + \quad C\equiv N\text{---}R^1 \quad + \quad Bu_3Sn\text{---}H \quad \xrightarrow{AIBN} \quad R\text{---}C\equiv N \quad + \quad R^1\text{---}H \quad + \quad Bu_3Sn\text{---}X$$

R = alkyl or aryl

isonitrile
$R^1 = {}^tBu, Bn$

$$\left[R^{\bullet} \quad + \quad {:}\overset{N-R^1}{\underset{\|}{C}} \quad \longrightarrow \quad R\text{---}\overset{N}{\underset{\text{imidoyl radical}}{C}}{}^{\bullet} \right]$$

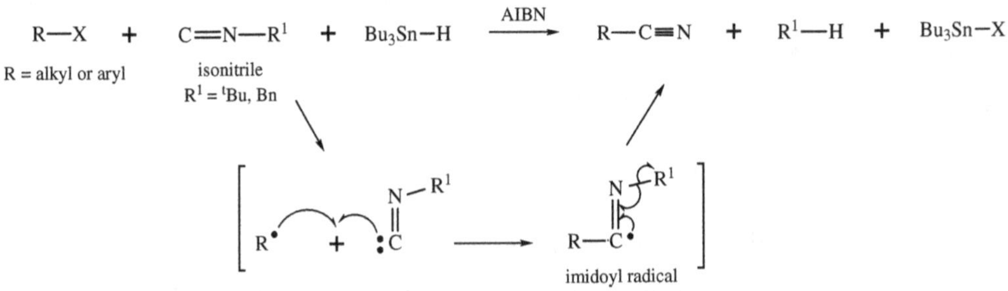

Fig. 6.26

However, a less hazardous and more general approach to nitriles has been developed which involves the addition of radicals to isonitriles (Fig. 6.26). Isonitriles, like carbon monoxide, possess a carbon atom with only six (rather than the preferred eight) outer electrons, and alkyl radicals can add to this carbon to form imidoyl radicals. These reactions are irreversible, but the imidoyl radicals can fragment to form the strong carbon–nitrogen triple bond of nitriles. The rate of β-fragmentation depends on the nature of the R^1 substituent and, when this is a *tert*-butyl or benzyl group, nitriles can be isolated in good yield.

The isonitrile approach can also be used to prepare aromatic nitriles, which cannot be formed by an S_N2 reaction, and this complements the Sandmeyer reaction of diazonium salts with copper(I) cyanide (Fig. 6.27).

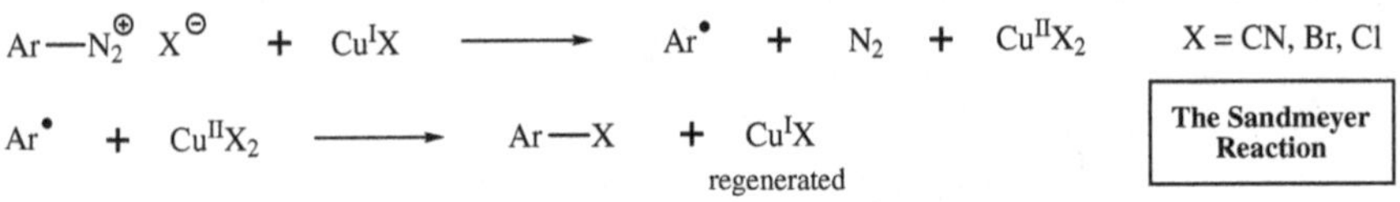

$$Ar\text{---}\overset{\oplus}{N_2} \; X^{\ominus} \quad + \quad Cu^IX \quad \longrightarrow \quad Ar^{\bullet} \quad + \quad N_2 \quad + \quad Cu^{II}X_2 \qquad X = CN, Br, Cl$$

$$Ar^{\bullet} \quad + \quad Cu^{II}X_2 \quad \longrightarrow \quad Ar\text{---}X \quad + \quad \underset{\text{regenerated}}{Cu^IX}$$

The Sandmeyer Reaction

Fig. 6.27

This reaction involves aryl radical intermediates produced by an electron transfer process, and other similar reactions can also be carried out using copper(I) halides. Thus, reaction of diazonium salts with copper(I) chloride or bromide leads to aryl chlorides or bromides, respectively. The reactions are terminated by ligand exchange from the copper ($Cu^{II}X_2$) to the benzene ring, and this is probably the best way of introducing a nitrile group, or bromine or chlorine atom, onto an aromatic ring.

6.2.7 Nitrosation

The conversion of an alkane (RH) to a nitroso compound (RNO) has proved to be synthetically very useful, and this can be carried out using radical intermediates. These reactions are very similar to alkane halogenations (see Section 6.2.1), and good selectivities can be achieved when using symmetrical alkane precursors (with only a single type of carbon–hydrogen bond).

This has been exploited in industry, and cyclohexane is photolysed with nitrosyl chloride to form cyclohexanone oxime by a non-chain radical mechanism (Fig. 6.28). This involves hydrogen atom abstraction by a chlorine atom, followed by reaction with nitric oxide. The nitroso product is in equilibrium (tautomerism) with the corresponding oxime, which is converted to caprolactam on treatment with acid (in a Beckmann rearrangement reaction). Caprolactam is subsequently converted to Nylon 6, and this short, economical approach has been used to produce 150 000 tonnes of caprolactam per year in the Toray process.

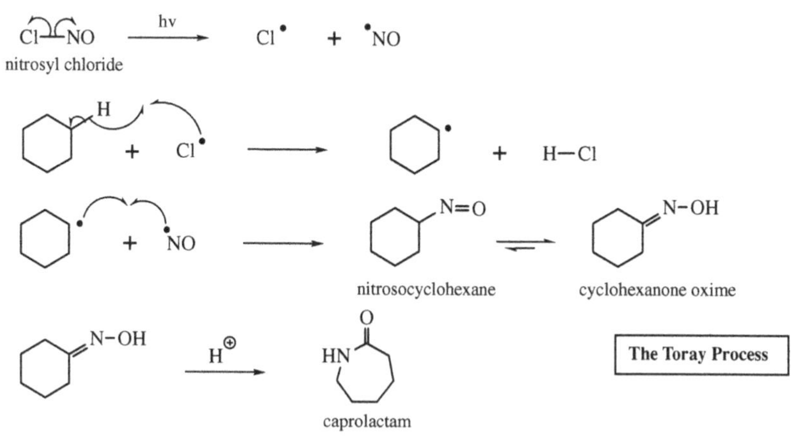

Fig. 6.28

137

6.3 Summary

Functional group transformations involving radical intermediates complement the many ionic transformations which are available. The mild reaction conditions and high selectivity, particularly for reduction reactions, are synthetically attractive. Radical deoxygenation and decarboxylation are particularly important because these are very general methods and can be used to reduce a variety of primary, secondary and tertiary alcohols and carboxylic acids. In contrast, the alternative ionic methods are much more limited. Oxidation and halogenations are generally more difficult to control and require activated precursors (with some bonds much weaker than others) or symmetrical starting materials for selective transformations.

Further reading

Crich, D. & Quintero, L. (1989) Radical chemistry associated with the thiocarbonyl group. *Chemical Reviews*, **89**, 1413–1432.

Fischer, M. (1978) Industrial applications of photochemical syntheses. *Angewandte Chemie, International Edition in English*, **17**, 16–26.

Hartwig, W. (1983) Modern methods for the radical deoxygenation of alcohols. *Tetrahedron*, **39**, 2609–2645.

Porter, N.A. (1986) Mechanisms for the autoxidation of polyunsaturated lipids. *Accounts of Chemical Research*, **19**, 262–268.

Ryu, I. & Sonoda, N. (1996) Free-radical carbonylations: then and now. *Angewandte Chemie, International Edition in English*, **35**, 1050–1066.

Ryu, I., Sonoda, N. & Curran, D.P. (1996) Tandem radical reactions of carbon monoxide, isonitriles, and other reagent equivalents of the geminal radical acceptor/radical precursor synthon. *Chemical Reviews*, **96**, 177–194.

Intramolecular Cyclization Reactions

7.1 Introduction

Some of the most important synthetic applications of radical chemistry involve intramolecular cyclization reactions. A variety of precursors can be selectively cyclized to afford both mono- and polycyclic products, generally in good yields. Radical cyclizations can give high levels of regio- and stereoselectivity, and a number of different methods can be used to carry out the cyclization reactions. All of these methods involve radical generation and cyclization (or cyclizations) onto a multiple bond, followed by reaction of the cyclized radical to form the product.

7.2 Ring size

Radical cyclizations are most often applied to the formation of five- and six-membered rings. These rings are relatively strain free, and are prepared via five- and six-membered transition states, which are easy to form. Cyclizations to form three- and four-membered rings are reversible, because of ring strain, and the equilibrium usually lies on the side of the acyclic radical precursor. However, cyclopropanes and cyclobutanes can be isolated if the cyclic radical (produced from a 3- or 4-*exo* cyclization) is stabilized by substituents, such as a nitrile group (Fig. 7.1). The presence of two (geminal) methyl groups restricts the conformation of the carbon chain and accelerates the cyclization by ensuring that the radical and alkene are in close proximity (this is known as the *gem*-dimethyl or Thorpe–Ingold effect).

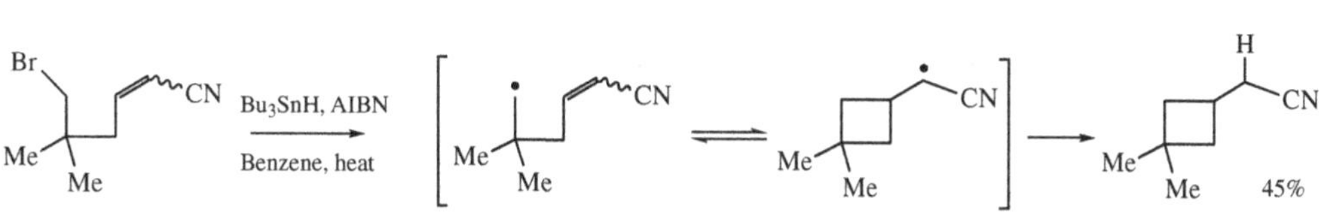

S-U. Park, T.R. Varick and M. Newcomb, *Tetrahedron Lett.*, 1990, **31**, 2975–2978

Fig. 7.1

For medium-sized rings (larger than six-membered), cyclization is difficult due to entropic factors; the reacting centres are far apart and so the radical is less likely to attack the alkene. There are, however, some examples of macrocyclization to form 10–20-membered rings or macrocycles (Fig. 7.2). These require very low concentrations of the starting material so as to minimize termination of the precursor radicals, and tributyltin hydride is generally added slowly to the precursor so as to minimize simple reduction (see Section 5.3.1). As for intermolecular addition reactions, the radical adds to the less hindered end of the double bond in an *endo* cyclization and, for good yields of cyclization, the radical and alkene acceptor must have opposite polarities, i.e. a nucleophilic radical will cyclize more rapidly onto an electron-poor double bond.

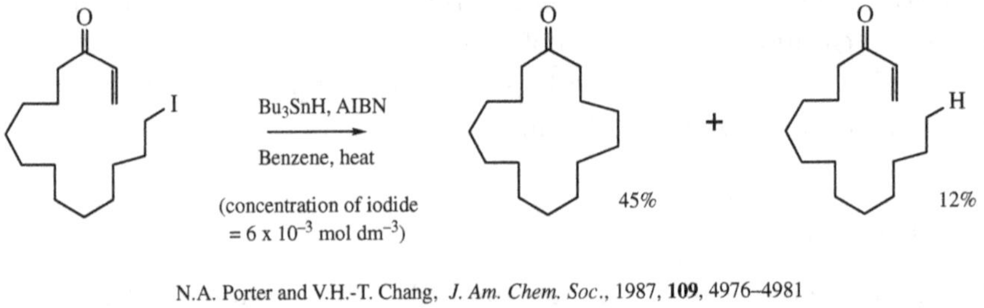

O
I
Bu_3SnH, AIBN
Benzene, heat
(concentration of iodide
= 6 x 10^{-3} mol dm^{-3})
O
45%
+
O
H
12%
N.A. Porter and V.H.-T. Chang, *J. Am. Chem. Soc.*, 1987, **109**, 4976–4981

Fig. 7.2

7.3 Regioselectivity

7.3.1 Hexenyl radical cyclization

Cyclopentyl rings are formed very rapidly on 5-*exo* cyclization of hex-5-en-1-yl radicals (Fig. 7.3). Reaction of bromide **(13)** with tributyltin hydride leads to an irreversible cyclization to form the primary radical **(14)** in preference to the thermodynamically more stable secondary radical **(15)**. The preference for 5-*exo* rather than 6-*endo* cyclization has been explained by stereoelectronic effects (see Section 4.7.4) favouring a chair-like transition state (sometimes called the 'Beckwith model') (Fig. 7.3).

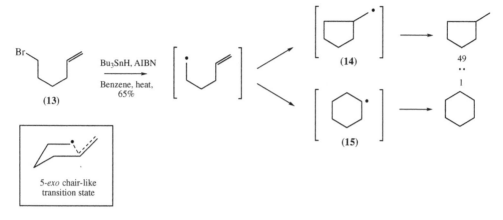

5-*exo* chair-like
transition state

Fig. 7.3

This mode of cyclization is observed for related systems and, although cyclizations onto alkyne triple bonds are slower, five-membered rings containing an exocyclic double bond are formed predominantly (Fig. 7.4). Cyclic radicals also mainly react in an *exo* manner with alkenes or alkynes to form bi-cycles, and *cis* fused ring products are preferred because these are less strained (Fig. 7.5).

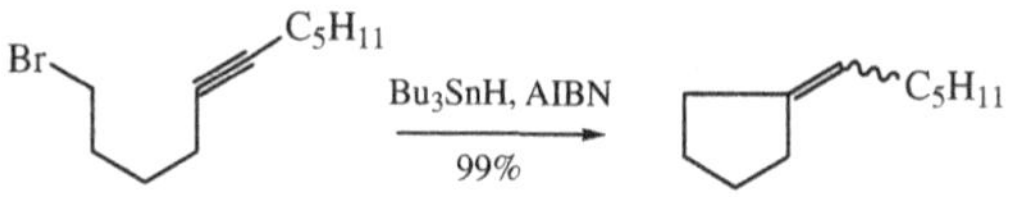

J.K. Crandell and D.J. Keyton,
Tetrahedron Lett., 1969, **21**, 1653–1656

Fig. 7.4

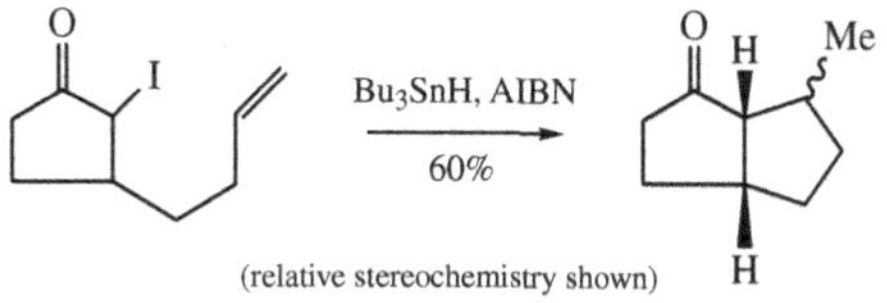

(relative stereochemistry shown)
C-K. Sha, T-S. Jean and D-C. Wang,
Tetrahedron Lett., 1990, **31**, 3745–3748

Fig. 7.5

The introduction of a heteroatom in the chain can increase both the regioselectivity and rate of 5-*exo* cyclization even further (Fig. 7.6). This is because the C—N—C and C—O—C bond angles (107.8° and 106.8°, respectively) are less than the tetrahedral C—C—C angle (109.5°), and the C—heteroatom bond length is shorter than for C—C (e.g. C—N is 1.47 Å, whereas C—C is 1.52 Å). These differences mean that the radical is closer to the internal carbon atom of the alkene, leading to better orbital overlap in the 5-*exo* (chair-like) transition state.

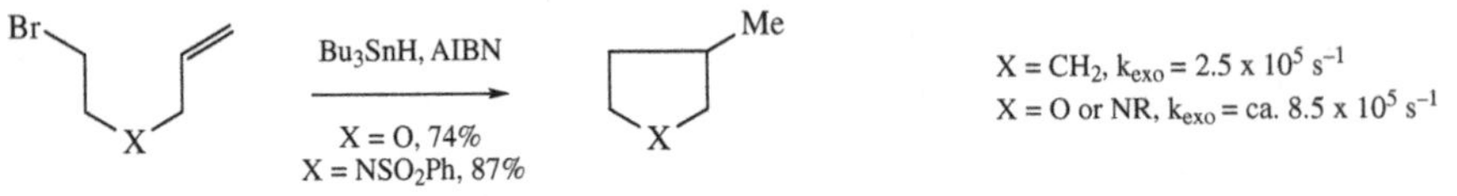
Br
X
Bu3SnH, AIBN
X = O, 74%
X = NSO2Ph, 87%
Me
X
X = CH2, kexo = 2.5 x 10^5 s^-1
X = O or NR, kexo = ca. 8.5 x 10^5 s^-1

Fig. 7.6

The rate and regioselectivity of cyclization are also influenced by the alkene substitution. Exclusive 5-*exo* cyclization is observed when substituents (R and R^1) are introduced at the terminus (or 6-position) of the alkene, because the formation of a six-membered ring would require attack at the sterically hindered carbon atom (Fig. 7.7). An electron-withdrawing substituent at the 6-position also accelerates the rate of 5-*exo* cyclization (of the nucleophilic radical) because of favourable frontier molecular orbital interactions (singly occupied molecular orbital–lowest unoccupied molecular orbital, SOMO–LUMO).

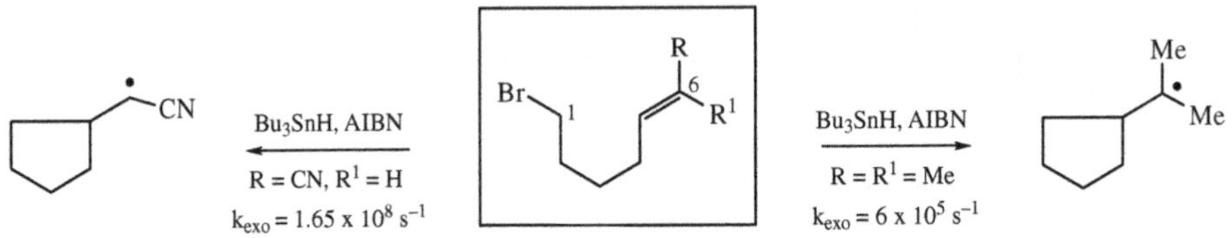
CN
Bu3SnH, AIBN
R = CN, R1 = H
kexo = 1.65 x 10^8 s^-1
R
Br
1
6
R1
Bu3SnH, AIBN
R = R1 = Me
kexo = 6 x 10^5 s^-1
Me
Me

Fig. 7.7

There are a few reactions for which the 6-*endo* product is preferred. When a substituent is introduced at the C-5 position, the rate of 5-*exo* cyclization is slowed down (due to steric effects) and the 6-*endo* cyclization can now effectively compete. For example, the introduction of a methyl substituent leads to twice as much of the six-membered ring product (Fig. 7.8) (cf. Fig. 7.3).

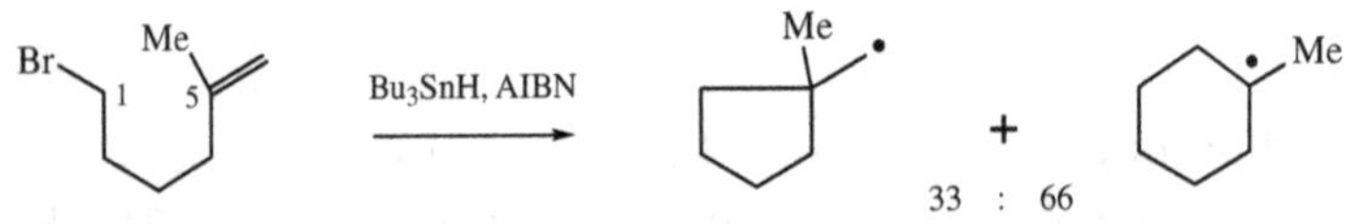
Br
Me
1
5
Bu3SnH, AIBN
Me
+
Me
33 : 66

Fig. 7.8

Six-membered rings are also isolated if the cyclization is reversible and under thermodynamic control. The first-formed radical needs to be stabilized so that ring opening of the 5-*exo* radical is faster than, for example, reduction (Fig. 7.9).

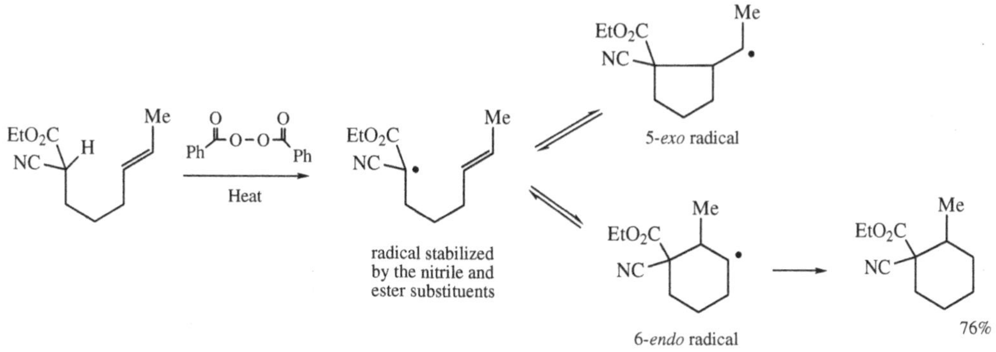

M. Julia, *Acc. Chem. Res.*, 1971, 386–392

Fig. 7.9

Reactive vinyl radicals generally give products derived from 5-*exo* cyclization (Fig. 7.10). Six-membered rings have been isolated from some reactions, although these are not usually derived from 6-*endo* cyclization, but from further reaction of the radical derived from 5-*exo* cyclization (Fig. 7.11). Slow addition of the tin hydride ensures that the intermediate methylenecyclopentyl radical **(16)** has time to undergo (reversible) 3-*exo* cyclization onto the alkene double bond, followed by ring opening to give the larger six-membered ring **(17)**. If the tin hydride is not added slowly, then radical **(16)** is reduced to give only the cyclopentane derivative **(18)**.

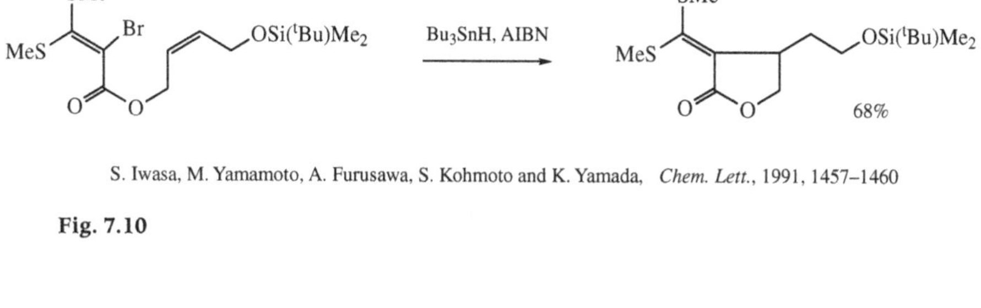

S. Iwasa, M. Yamamoto, A. Furusawa, S. Kohmoto and K. Yamada, *Chem. Lett.*, 1991, 1457–1460

Fig. 7.10

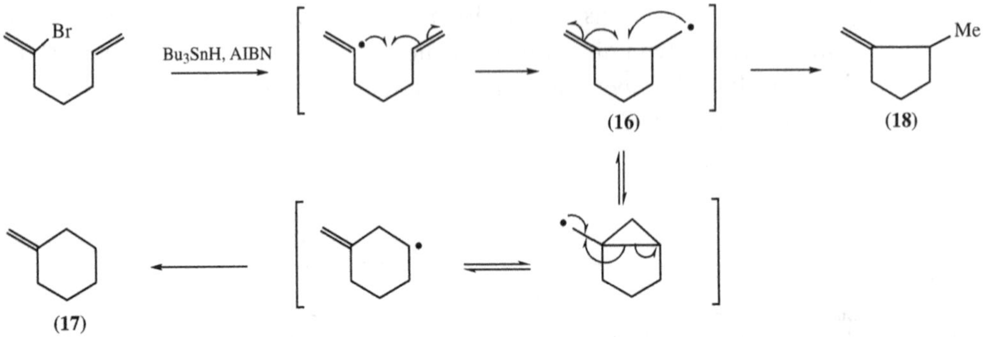

(16)   (18)

(17)

Fig. 7.11

Aryl radicals generally cyclize to give five-membered rings (Fig. 7.12). For example, reaction of aryl iodide **(19)** with samarium(II) iodide produces an aryl radical that undergoes exclusive 5-*exo* cyclization to give a primary radical. This is subsequently reduced to form an organosamarium(III) species, which can add to a ketone, for example, to give tertiary alcohol **(20)** (after work-up).

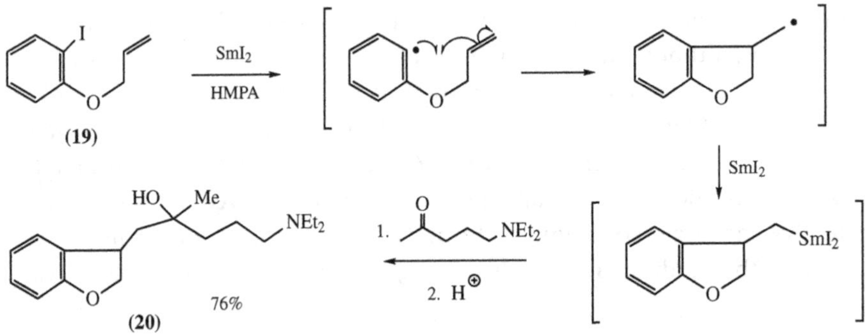

D.P. Curran, T.L. Fevig and M.J. Totleben, *Synlett*, 1990, 773–774

Fig. 7.12

Aryl radicals, like vinyl radicals, can undergo a ring expansion reaction to form a six-membered ring, particularly when an activating substituent (such as a carbonyl group) is present on the benzene ring (Fig. 7.13). This stabilizes the radical derived from 3-*exo* cyclization and a low concentration of tin hydride allows the six-membered ring product, as well as the five-membered ring product, to be formed.

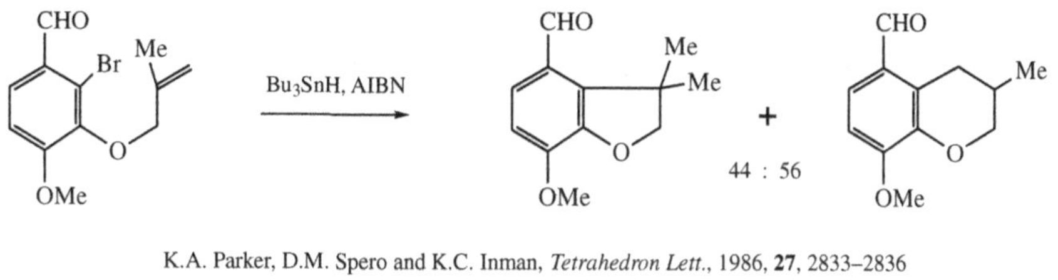

K.A. Parker, D.M. Spero and K.C. Inman, *Tetrahedron Lett.*, 1986, **27**, 2833–2836

Fig. 7.13

7.3.2 Heptenyl radical cyclization

The hept-6-en-1-yl radical also prefers to undergo *exo* cyclization to give predominantly the cyclohexane product rather than cycloheptane product via a chair-like transition state (Fig. 7.14). However, the cyclization is around 40 times slower than the hex-5-en-1-yl cyclization, and competing processes, including simple reduction and particularly 1,5-hydrogen atom transfer, are more important in 6-*exo* cyclization reactions. In the 1,5-hydrogen atom transfer, the primary radical is able to abstract the hydrogen atom in a six-membered transition state to generate a resonance-stabilized allylic radical.

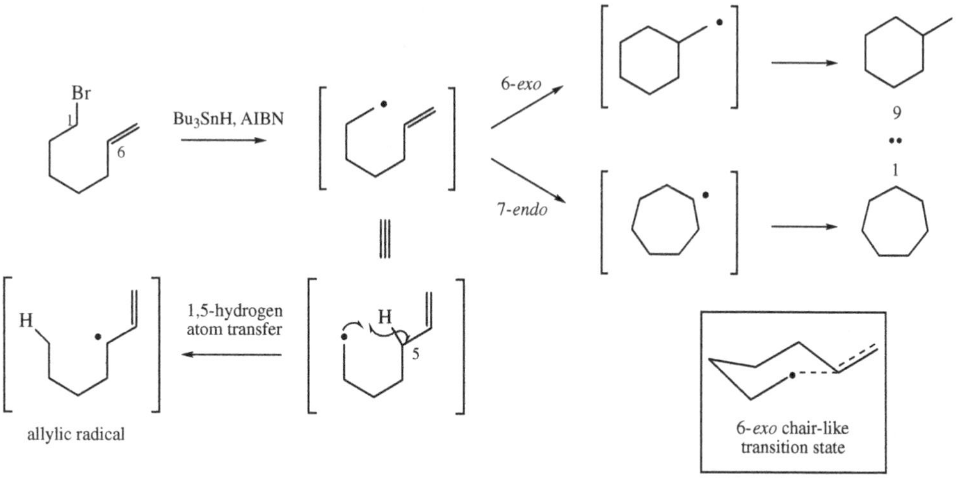

Fig. 7.14

This is usually minimized by introducing electron-withdrawing substituents onto the alkene (at position 7), which increases the rate of 6-*exo* cyclization because of a favourable SOMO–LUMO interaction. Thus, the primary radical derived from bromide **(21)** efficiently cyclizes onto the electron-poor double bond to give cyclohexane **(22)** (as a 1.2 : 1 mixture of diastereoisomers) in an excellent 91% yield (Fig. 7.15).

Me$_2$(^{t}Bu)SiO,,,,

Br   CO$_2$Me

→ Bu$_3$SnH, AIBN →

Me$_2$(^{t}Bu)SiO,,,,   CO$_2$Me

OBn   OBn   91%

(21)   **(22)**

J. Marco-Contelles and B. Sanchez, *J Org. Chem.*, 1993, **58**, 4293–4297

Fig. 7.15

7.4 Stereoselectivity

7.4.1 Hexenyl radical cyclization

The chair-like transition state model for 5-*exo* cyclization provides a rationale for the stereoselectivities observed on cyclization of substituted hexenyl radicals. Substituents at C-2, C-3 and C-4 would be expected to adopt a pseudo-equatorial position so as to avoid unfavourable steric (1,3-diaxial) interactions (Fig. 7.16). The larger the substituent, the stronger the axial interactions, and therefore the greater the proportion of the equatorial conformer. This explains why 3-substituted radicals preferentially give the *cis*-disubstituted cyclopentyl products, while 2- or 4-substituted radicals preferentially give the *trans*-disubstituted cyclopentyl products (Fig. 7.17).

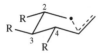

Fig. 7.16

The chair-like transition state model also explains why *cis*-disubstituted cyclopentyl products are formed preferentially on cyclization of 1-

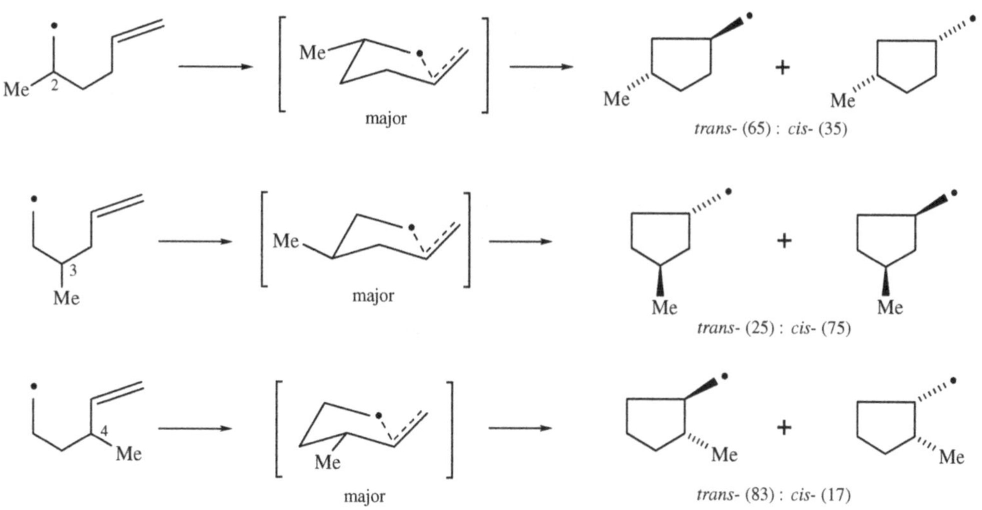

Fig. 7.17

substituted radicals (Fig. 7.18). In the transition state, the C-1 substituent appears to be in close proximity to the methylene group of the alkene, but is in fact also equatorial in the cyclohexane transition state and is therefore in the less sterically hindered position. This is supported by modelling calculations which show that the steric interaction is small and that the alternative transition state, with the C-1 substituent in an axial position, is higher in energy.

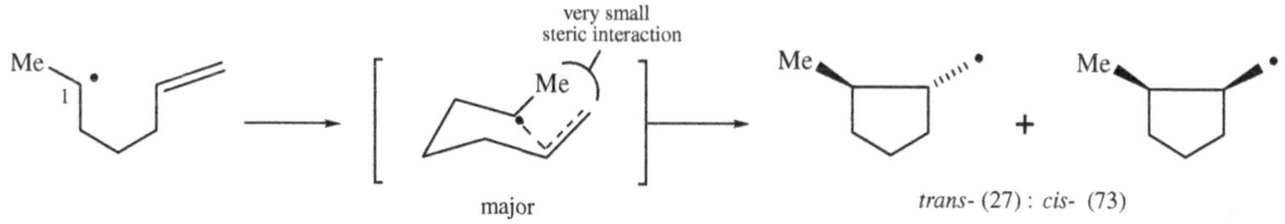

Fig. 7.18

This explains why radical **(24)**, derived from oxymercuration of enol ether **(23)**, cyclizes onto the α,β-unsaturated ester to produce predominantly the *cis*- rather than the *trans*-disubstituted cyclopentane **(25)** (Fig. 7.19).

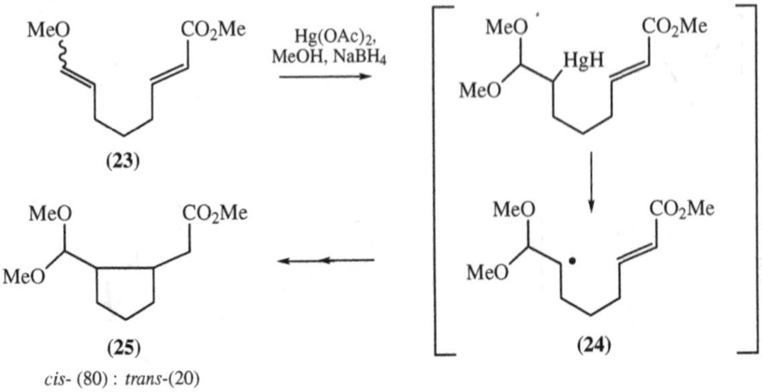

K. Weinges and W. Sipos, *Chem. Ber.*, 1988, **121**, 363–368

Fig. 7.19

This model can be extended to the cyclization of trisubstituted precursors (Fig. 7.20). For example, reaction of primary bromide **(26)** with tributyltin hydride produces the trisubstituted tetrahydrofuran as a mixture of two diastereoisomers (in a ratio of 7 : 1). The stereochemistry of the major isomer **(27)** is that predicted from a chair-like transition state.

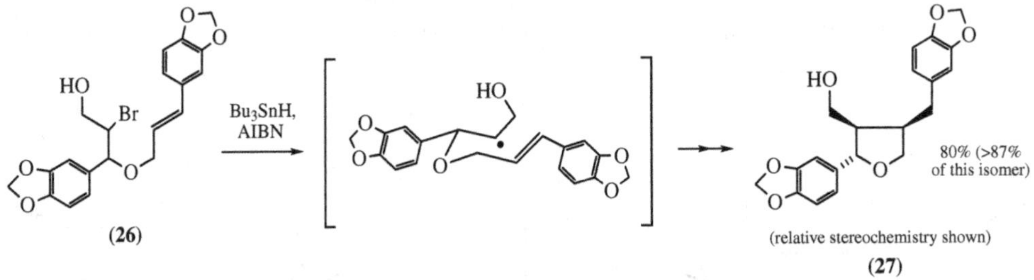

G. Maiti, S. Adhikari and S.C. Roy, *J. Chem. Soc., Perkin Trans. 1*, 1995, 927–929

Fig. 7.20

7.4.2 Heptenyl radical cyclization

The chair-like transition state model can also be used to explain the stereoselectivity observed on 6-*exo* cyclization of the hept-6-en-1-yl radical (Fig. 7.21). The substituent and alkene acceptor prefer to adopt pseudoequatorial positions, and 2- or 4-substituted radicals give preferentially *cis*-disubstituted cyclohexyl products, whilst 1-, 3- or 5-substituted radicals preferentially give the *trans*-disubstituted cyclohexyl products.

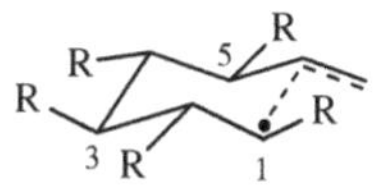

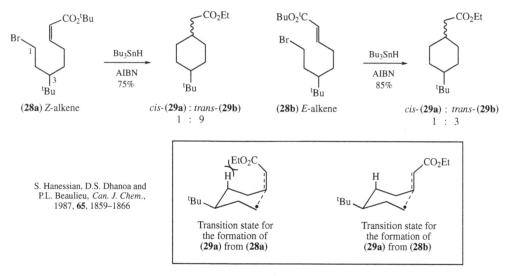

Fig. 7.21

This is illustrated in the stereoselective cyclization of bromides **(28a)** and **(28b)** (which have a substituent at the 3-position), which gives predominantly the *trans*-disubstituted cyclohexane **(29b)** (Fig. 7.22). The *Z*-alkene **(28a)** gives a better stereoselectivity than the *E*-alkene **(28b)**, and this can be explained by a comparison of the two different transition states leading to *cis*-cyclohexane **(29a)**. In both cases, the ester substituent must be in the less favourable axial position, and whereas the *Z*-ester substituent interacts with the axial hydrogen to raise the energy of this transition state, this is not possible in the *E*-ester transition state. Cyclization of the *Z*-ester **(28a)** therefore leads to a greater preference for the *trans*-cyclohexane **(29b)**.

S. Hanessian, D.S. Dhanoa and
P.L. Beaulieu, *Can. J. Chem.*,
1987, **65**, 1859–1866

Fig. 7.22

7.5 Cyclization reactions leading to rearrangement

Radical cyclization reactions can be used to generate reactive radicals that undergo a subsequent fragmentation or β-elimination reaction to give, overall, a radical rearrangement. The 1,2-shift of aromatic rings is an important example of a radical rearrangement, and this is initiated by cyclization onto a benzene ring (see Section 4.5).

A related and synthetically useful reaction is the 'Beckwith–Dowd' ring expansion of cycloalkanones (Fig. 7.23). Cyclization of a primary radical onto the carbonyl double bond is reversible, and the intermediate alkoxyl radical can fragment to regenerate the starting radical or, alternatively, form a more stable tertiary and resonance-stabilized radical, which leads to a one-carbon increase in the size of the ring. The ester substituent effectively stabilizes the cyclohexyl radical, and this is a general method which can be applied to a wide range of substrates; examples of one- to four-carbon ring expansions are known.

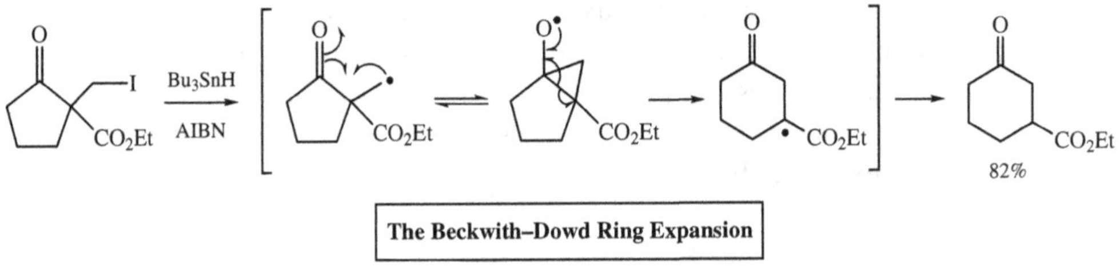

The Beckwith–Dowd Ring Expansion

Fig. 7.23

7.6 Synthetic applications

Some examples of radical cyclizations will now be discussed in more detail to show the importance of these types of reaction in the synthesis of natural products.

A particularly elegant synthesis of the terpene hirsutene (**34**) has been developed using a tandem radical cyclization reaction as the key step (Fig. 7.24). The ability to form more than one carbon–carbon bond (or ring) in a single reaction is synthetically very attractive, and this approach was used to form the tri-cyclic ring system of racemic hirsutene (**34**) from the iodide (**30**).

Reaction of (**30**) with the tributyltin radical, prepared from tributyltin hydride, gives the primary radical (**31**); this undergoes a favoured 5-*exo* cyclization to produce the bi-cyclic radical (**32**). The newly formed ring junction was exclusively *cis* because the short chain connecting the radical

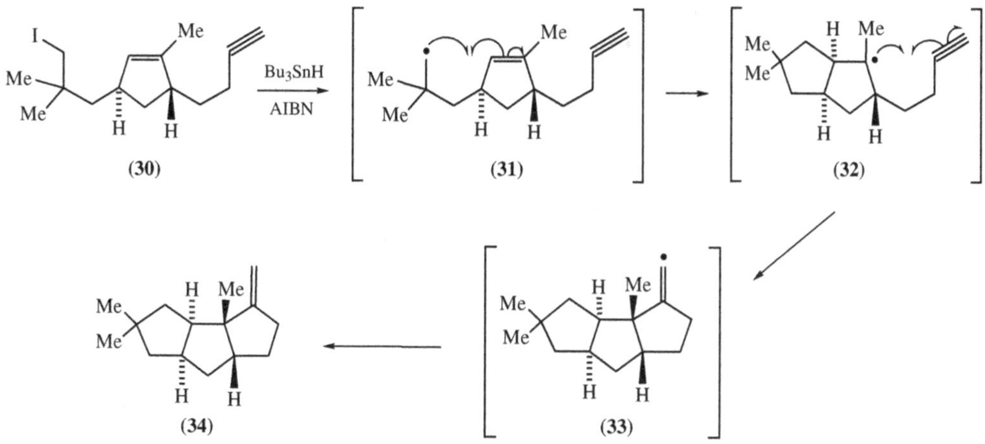

D.P. Curran and D.M. Rakiewicz, *J. Am. Chem. Soc.*, 1985, **107**, 1448–1449

Fig. 7.24

and ring ensured that the alkene could only be attacked from the 'top face'. Any attack from the bottom face would require considerable ring strain. The tertiary radical **(32)** was then able to undergo a slower 5-*exo* cyclization onto the alkyne triple bond to give the reactive vinyl radical **(33)**. This, like the first cyclization, is thermodynamically favoured because a σ bond is formed at the expense of an alkyne π bond, and the stereoselective formation of a second *cis* fused ring system is also due to ring strain. Finally, the vinyl radical **(33)** is reduced by reaction with tin hydride to give hirsutene **(34)** in approximately 80% yield. The use of iodide **(30)** with *trans*-disposed side chains therefore ensures that the correct ring fusion is obtained for the natural product. This synthesis shows how radical intermediates can be used to construct hindered carbon–carbon bonds and quaternary centres in excellent yield.

Related terpenes can be prepared using this type of tandem cyclization sequence. For example, a samarium(II) iodide-mediated cyclization of aldehyde **(35)** has been used to prepare, after hydrolysis of the acetal, the tri-cyclic keto alcohol **(37)** in 60% yield (Fig. 7.25). Addition of hexamethylphosphoramide (HMPA) was found to accelerate the cyclization reaction, and the vinyl radical **(36)** was thought to abstract a hydrogen atom from the solvent (tetrahydrofuran, THF) rather than undergo further reduction to a vinyl anion [which would require at least 2 equivalents of samarium(II)]. Oxidation of (37), to selectively introduce an epoxide group, completed the synthesis of the terpene hypnophilin (38).

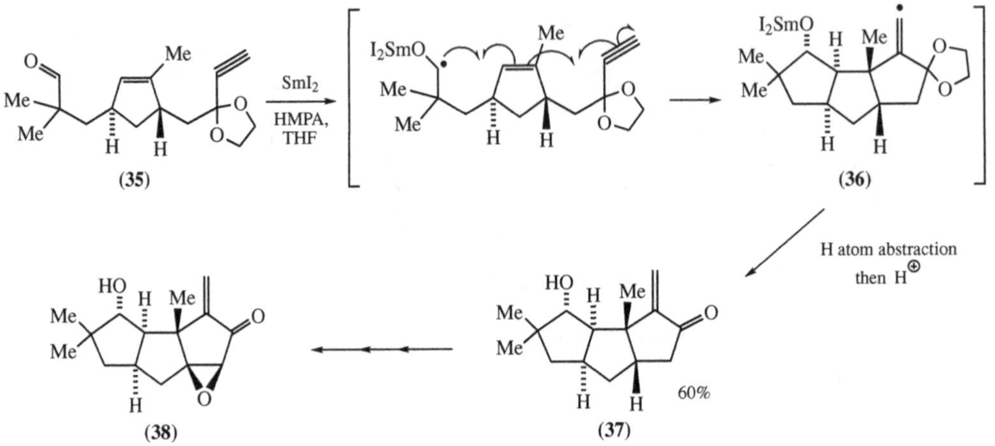

T.L. Fevig, R.L. Elliott and D.P. Curran, *J. Am. Chem. Soc.*, 1988, **110**, 5064–5067

Fig. 7.25

A similar tandem radical cyclization strategy has been used to prepare the important alkaloid morphine (39) (Fig. 7.26). This molecule has attracted considerable synthetic attention because of its challenging structure and important medicinal properties as a pain suppressant.

Reaction of aryl bromide (40) with tributyltin hydride chemoselectively breaks the carbon–bromine bond to generate the reactive aryl radical (41), which undergoes 5-*exo* cyclization onto the trisubstituted alkene to give (42). A five- rather than six-membered ring is preferentially formed because, although addition takes place on the more substituted alkene carbon, this is held closer to the attacking radical. As in the hirsutene (34) synthesis, a stable (less strained) *cis* fused ring product is formed. The resulting cyclic secondary radical (42) then undergoes a 6-*endo* (rather than 5-*exo*) cycliza-

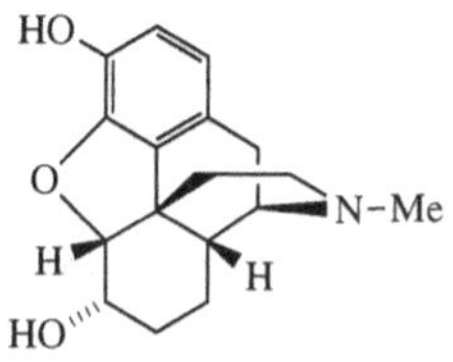

Morphine (39)

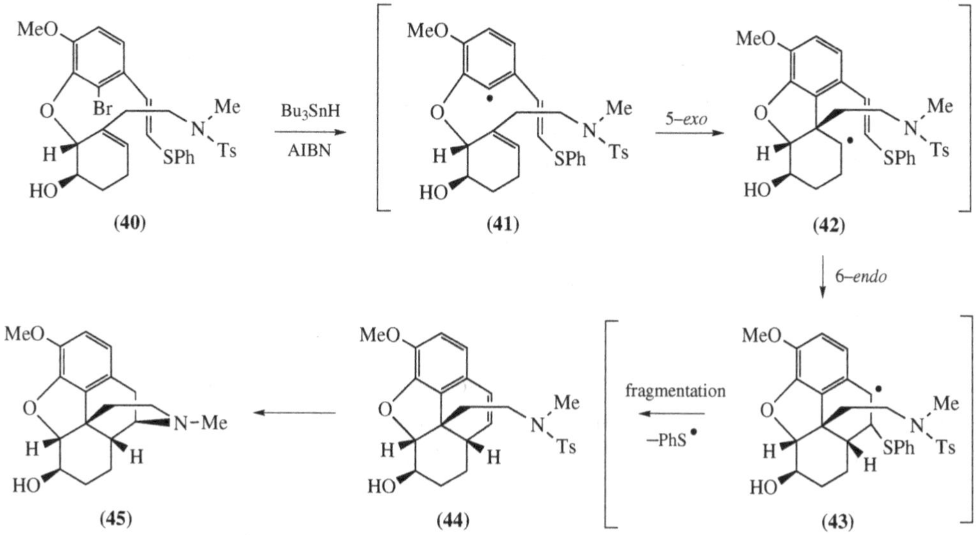

K.A. Parker and D. Fokas, *J. Am. Chem. Soc.*, 1992, **114**, 9688–9689

Fig. 7.26

tion to form the tetracyclic radical (43). This mode of cyclization produces a less strained ring system than that derived from the alternative 5-*exo* cyclization, and also generates a resonance-stabilized benzylic radical. Finally, rapid fragmentation of radical (43) occurs to break the weak C—S bond and form alkene (44); the thiyl radical (PhS•) is a good leaving group as the radical can be stabilized by delocalization. The fragmentation introduces an alkene double bond, which is required for a subsequent cyclization reaction. Although the tetracyclic product (44) is isolated in only 35% yield from (40), this radical reaction has accomplished two cyclizations and an elimination in one reaction. The mild reaction conditions are also compatible with the alcohol functional group and this does not need to be protected before cyclization. To complete the synthesis, the N-tosyl protecting group is removed and, under the reducing reaction conditions, the nitrogen spontaneously reacts with the double bond to form dihydroisocodeine (45), which can be readily converted to morphine (39).

This laboratory synthesis has some similarities to the way in which morphine (39) is prepared in nature (Fig. 7.27). The biosynthesis is also believed to involve a radical cyclization, known as an oxidative phenolic coupling reaction. Morphine (39) is known to be derived from (R)-reticuline (46), which contains two phenol rings, and these can be oxidized to generate diradical (47). The phenol oxygens may lose an electron to give radical cations that can form radicals on deprotonation (see Chapter 11). These radicals can then couple intramolecularly to form a new carbon–carbon bond at the 2- and 4-positions of the two phenol rings; this is often known as an *ortho–para* coupling reaction. The product salutaridine (48) is transformed into morphine (39) by a series of ionic reactions, which includes a cyclization reaction to form the remaining five-membered ring. Like the laboratory synthesis, the structurally challenging stereogenic quaternary centre (at position 4 in Fig. 7.27) of morphine (39) is introduced by a radical cyclization reaction.

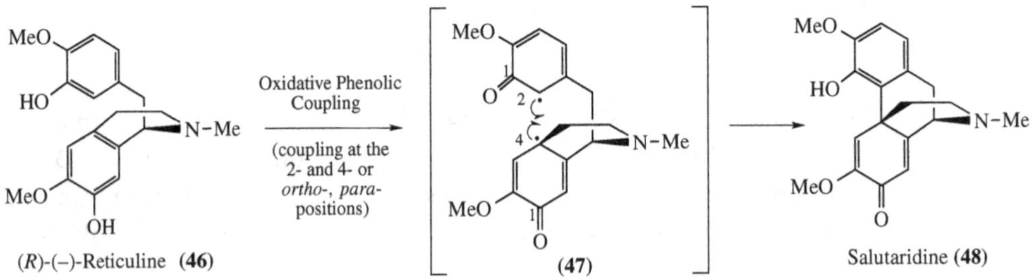

Fig. 7.27

In contrast to tributyltin hydride-mediated reactions, an atom transfer method can allow the introduction of a versatile halogen atom into the cyclized product. This has been exploited in a racemic synthesis of the alkaloid haemanthidine (49) (Fig. 7.28).

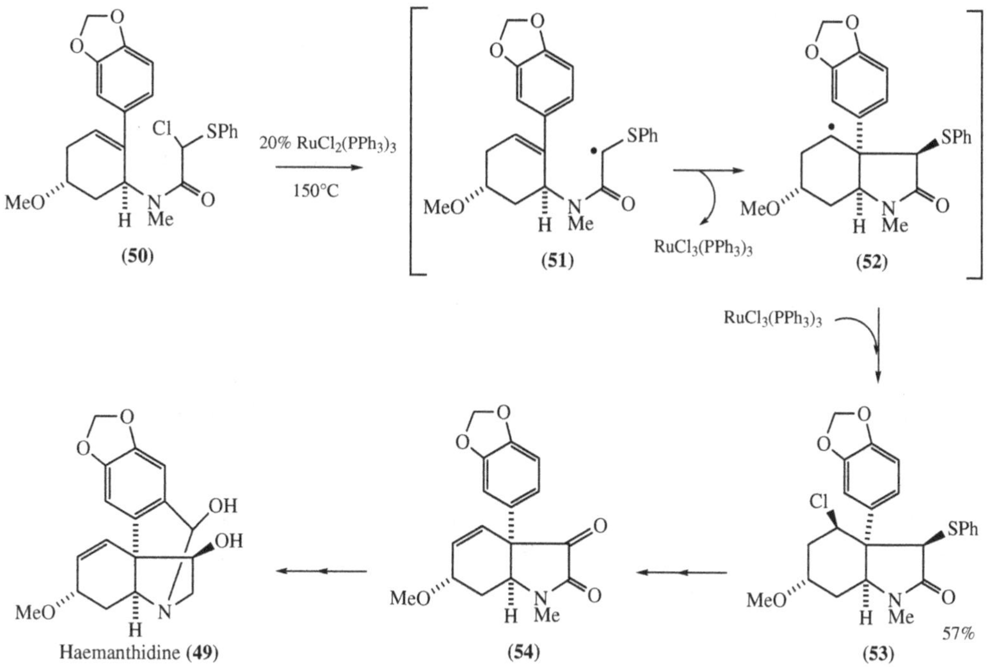

H. Ishibashi, N. Uemura, H. Nakatani, M. Okazaki, T. Sato, N. Nakamura and M. Ikeda, *J. Org. Chem.*, 1993, **58**, 2360–2368

Fig. 7.28

The key step involves heating chloro sulfide **(50)** with a ruthenium(II) chloride complex to generate radical **(51)** [together with ruthenium(III)]. This undergoes a 5-*exo* cyclization to give radical **(52)**, which is more reactive than **(51)**, and so can react with the ruthenium(III) complex to introduce a chlorine atom on the cyclohexane ring of **(53)**. In the process, ruthenium(II) is regenerated and so less than 1 equivalent of the initiator can be used. A stable *cis* fused bi-cyclic system is formed and the aromatic side chains on the five-membered ring of **(53)** have a *trans* arrangement, presumably so as to minimize steric interactions. The chlorine atom is also introduced stereoselectively from the top face because the aromatic ring (at the ring junction) effectively shields the bottom face of the molecule from attack. The introduction of a chlorine atom is crucial for the synthesis as the natural product **(49)** contains a cyclohexene ring, and **(53)** can be converted to the alkene **(54)** in a number of steps, including elimination of HCl to introduce the double bond. Alkene **(54)** can then be elaborated to haemanthidine **(49)**.

Radical cyclizations to form alkenes can be carried out in one step using cobalt(I) initiators, and this has been applied to the synthesis of kainic acid

(55) (Fig. 7.29). When iodide **(56)** is treated with a cobalt(I) reagent, nucleophilic substitution produces the cobalt(III) species **(57)**. Homolysis of the cobalt–carbon bond generates cobalt(II) together with a carbon-centred radical **(58)**, and this can cyclize to form the five-membered ring **(59)**. The chiral centre (at the C-2 position) controls the stereochemistry at the 3-position and steric interactions ensure that these side chains are *trans* in cyclic radical **(59)**. In addition, a predominant *cis* relationship between the substituents at the 3- and 4-positions is consistent with that expected for cyclization of 1-substituted hexenyl radicals (see Section 7.4.1). The cyclic radical **(59)** adds to cobalt(II) to form a new cobalt(III) species **(60)** that can undergo dehydrocobaltation to give, predominantly, the less substituted alkene **(61)**. The preference for the less substituted alkene is believed to be due to steric effects, i.e. the transition state (*syn* elimination) leading to the terminal alkene is less strained and so more easily formed. Overall, this cyclization has produced a substituted ring with the correct stereochemistry and alkenic side chain required for the synthesis of kainic acid **(55)**.

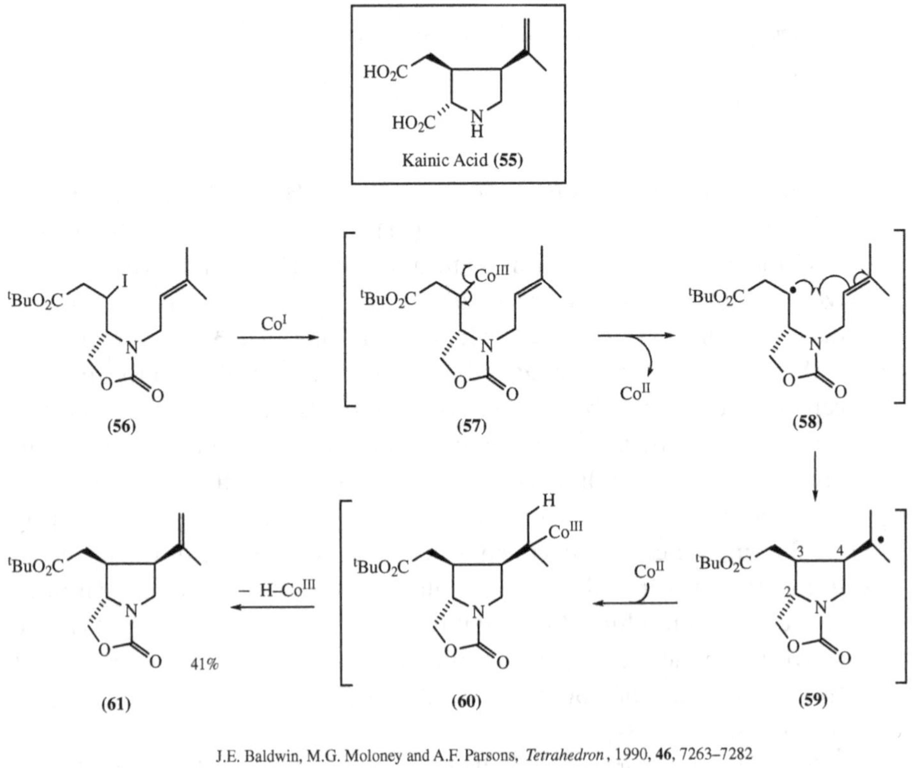

Kainic Acid **(55)**

J.E. Baldwin, M.G. Moloney and A.F. Parsons, *Tetrahedron*, 1990, **46**, 7263–7282

Fig. 7.29

Alternatively, alkenic products can be derived from reaction of carbonyl compounds with manganese(III) and copper(II) reagents. This type of reaction has been used to prepare avenaciolide (62) from the α-chloro-diester (63) (Fig. 7.30). Reaction of (63) with manganese(III) acetate generates the radical at the α-position, and (64) undergoes intramolecular 5-*exo* cyclization to give the secondary radical (65). This radical can be oxidized by an equivalent of copper(II) acetate to produce an organocopper(III) intermediate (66), which undergoes an oxidative elimination to form the less substituted alkene (67). The stereochemistry at the C-4 position is controlled by the octyl substituent at position 5, and steric interactions ensure that these groups are *trans* to each other in (65)–(67). A chloro substituent at the α-position of (67) is crucial to the synthesis as this prevents any further radical generation at this position—the corresponding product with an α-hydrogen substituent was shown to undergo rapid oxidation leading to decomposition. Chloroalkene (67) was elaborated to avenaciolide (62) and this involved displacement of the chlorine substituent [in an S_N2 reaction (substitution, nucleophilic, bimolecular)] to form the second lactone ring.

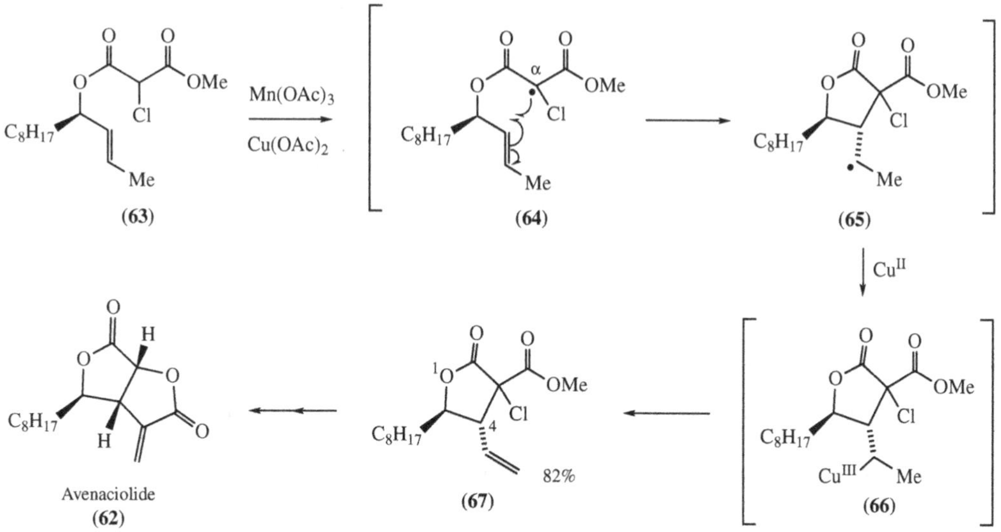

B.B. Snider and B.A. McCarthy, *Tetrahedron*, 1993, **49**, 9447–9452

Fig. 7.30

Radical cyclization reactions can be used to prepare large rings (or macrocycles) which contain 10 or more atoms, as well as the more common five- and six-membered ring targets. This is illustrated by the synthesis of (−)-zearalenone (68) (which contains a 14-membered ring) from reaction of bromide (69) with tris(trimethylsilyl)silane (Fig. 7.31). Bromine atom abstraction generates a nucleophilic allylic radical which, under high dilution conditions, cyclizes onto the (electron-poor) α,β-unsaturated double bond in an *endo* manner to form the 14-membered ring. The cyclization gives ketone (70) in 55% yield and subsequent cleavage of the two methyl ethers produces the natural product (68).

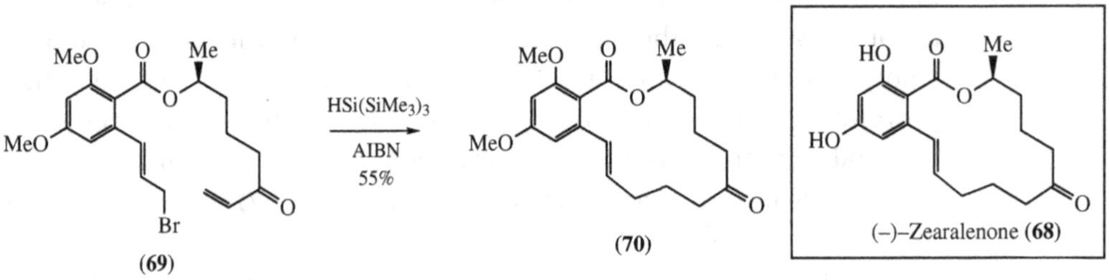

S.A. Hitchcock and G. Pattenden, *J. Chem. Soc., Perkin Trans. 1*, 1992, 1323-1328

Fig. 7.31

7.7 Summary

Radical cyclization reactions are ideally suited to the formation of five- and six-membered rings. The reactions mainly lead to products of kinetic control and even complex polycyclic molecules can be prepared, often in one pot, under mild and neutral reaction conditions that are tolerant to a variety of functional groups.

Further reading

Beckwith, A.L.J. & Schiesser, C.H. (1985) Regio- and stereo-selectivity of alkenyl radical ring closure: a theoretical study. *Tetrahedron*, **41**, 3925–3941.

Bowman, W.R., Bridge, C.F. & Brookes, P. (2000) Synthesis of heterocycles by radical cyclisation. *Journal of the Chemical Society, Perkin Transactions 1*, **1**, 1–14.

Bunce, R.A. (1995) Recent advances in the use of tandem reactions for organic synthesis. *Tetrahedron*, **51**, 13103–13159.

Dowd, P. & Zhang, W. (1993) Free radical-mediated ring expansion and related annulations. *Chemical Reviews*, **93**, 2091–2115.

Fallis, A.G. & Brinza, I.M. (1997) Free radical cyclizations involving nitrogen. *Tetrahedron*, **53**, 17543–17594.

Giese, B., Kopping, B., Göbel, T., Dickhaut, J., Thoma, G., Kulicke, K.J. & Trach, F. (1996) Radical cyclization reactions. *Organic Reactions*, **48**, 301–856.

Julia, M. (1971) Free-radical cyclizations. *Accounts of Chemical Research*, **4**, 386–392.

Parsons, P.J., Penkett, C.S. & Shell, A.J. (1996) Tandem reactions in organic synthesis: novel strategies for natural product elaboration and the development of new synthetic methodology. *Chemical Reviews*, **96**, 195–206.

Spellmeyer, D.C. & Houk, K.N. (1987) A force-field model for intramolecular radical additions. *Journal of Organic Chemistry*, **52**, 959–974.

Yet, L. (1999) Free radicals in the synthesis of medium-sized rings. *Tetrahedron*, **55**, 9349–9403.

CHAPTER 8

Intermolecular Reactions

8.1 Introduction

Intermolecular reactions are much more difficult to carry out than comparable cyclizations because of entropic factors—it is much easier for a molecule to react with 'itself' than with another molecule. The most common intermolecular reactions involve the regioselective addition of a radical to an alkene or alkyne (Fig. 8.1). In order that the radical **(71)** can add to the alkene before dimerization, for example, it is often vital that the polarity of the radical is matched to that of the double bond to ensure a fast rate of addition. For reactions mediated by tributyltin hydride, the addition reaction must also compete favourably with simple reduction (leading to R—H). The use of an excess of the acceptor alkene (or alkyne) can increase the rate of radical addition, but this can also lead to polymerization. Therefore, if polymerization is to be avoided, radical **(72)** must abstract a hydrogen atom from tributyltin hydride before addition to another molecule of alkene can take place. For a good yield of **(73)**, the tributyltin hydride must therefore react quickly with radical **(72)**, but only slowly with radical **(71)**, and this can be achieved if the radicals have different polarities and hence different substituents.

Desired Reactions

$$R^\bullet \ + \ CH_2{=}CH{-}R^1 \ \longrightarrow \ R{-}CH_2{-}\overset{\bullet}{C}H{-}R^1$$
(71) (72)

$$R{-}CH_2{-}\overset{\bullet}{C}H{-}R^1 \ + \ Bu_3Sn{-}H \ \longrightarrow \ R{-}CH_2{-}CH_2{-}R^1 \ + \ Bu_3Sn^\bullet$$
(73)

Competing Reactions

Simple Reduction

$$R^\bullet \ + \ Bu_3Sn{-}H \ \longrightarrow \ R{-}H \ + \ Bu_3Sn^\bullet$$
(71)

Polymerization

$$R{-}CH_2{-}\overset{\bullet}{C}H{-}R^1 \ + \ CH_2{=}CH{-}R^1 \ \longrightarrow \ R{-}CH_2{-}CH(R^1){-}CH_2{-}\overset{\bullet}{C}H{-}R^1 \ \longrightarrow \ Polymer$$
(72)

Fig. 8.1

8.2 Alkene substitution

In Section 4.3.1, the regioselective addition of a radical to an unsymmetrical alkene was discussed, and the radical was shown to add to the less substituted end of the double bond (because this is less crowded and more accessible). The size of the alkene substituents will therefore influence the regioselectivity and a radical will tend to add to the (alkene) carbon atom having the smaller substituents. In addition, the electronic (or polar) effects of the alkene substituents are very important. For nucleophilic radicals, the alkene should have an electron-withdrawing group (EWG) to increase the singly occupied molecular orbital–lowest unoccupied molecular orbital (SOMO–LUMO) interaction, whilst, for electrophilic radicals, the alkene should have an electron-donating group to increase the singly occupied molecular orbital–highest occupied molecular orbital (SOMO–HOMO) interaction.

Alkyl radicals which are nucleophilic will therefore react relatively quickly (with rate constants of $\approx 10^5$–10^6 dm^3 mol^{-1} s^{-1} at 20°C) with electrophilic alkenes, and the cyclohexyl radical, for example, can add very efficiently to a variety of alkenes bearing EWGs (Fig. 8.2). In these reactions, the tributyltin hydride is generated *in situ* by reduction of only 0.2 equivalent of tributyltin chloride; this ensures a low concentration of tributyltin hydride, which minimizes simple reduction (to form cyclohexane). The alkene acceptor is used in excess to increase the rate of radical addition, and these reactions work well because the radical intermediates have different polarities. Addition of the cyclohexyl radical to electron-poor alkenes produces intermediate secondary radicals, which have an adjacent EWG and therefore are much less nucleophilic than the cyclohexyl radical. This means that addition to another molecule of alkene, leading to polymerization, is much slower, and therefore the intermediate radical can be trapped (by reaction with tributyltin hydride) before addition can take place.

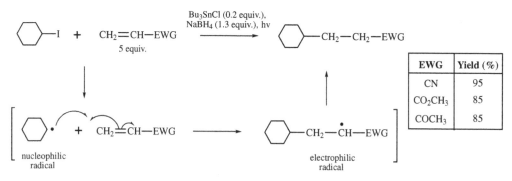

EWG	Yield (%)
CN	95
CO$_2$CH$_3$	85
COCH$_3$	85

B. Giese, J.A. Gonzales-Gomez and T. Witzel, *Angew. Chem.*, 1984, **96**, 51–52

Fig. 8.2

Similar reactions can be carried out using other initiators (Fig. 8.3). For example, reaction of cyclohexylmercury(II) acetate with sodium borohydride and methyl vinyl ketone gives the product of intermolecular addition in 77% yield. The yield is lower than that obtained from the corresponding tin hydride reaction, because the intermediate mercury hydride ($C_6H_{11}HgH$) reacts at least 10 times faster with alkyl radicals (than does Bu_3SnH), and so this leads to more simple reduction (to give cyclohexane).

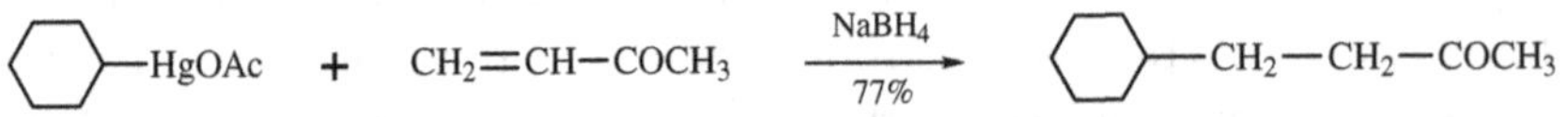

B. Giese and G. Kretzschmar, *Chem. Ber.*, 1983, **116**, 3267–3270

Fig. 8.3

The Meerwein arylation reaction employs a similar intermolecular addition using aryl radicals generated from reaction of a diazonium salt with a catalytic amount of copper(I) chloride (Fig. 8.4). The reactive phenyl radical (which is nucleophilic) typically adds to an electron-poor alkene, such as acrylonitrile, to give a radical that is converted to the secondary chloride **(74)** on reaction with copper(II) chloride. Reaction of copper(II) chloride with other aryl radicals and alkenes can produce alkene products **(75)**. These can be formed by oxidation of the radical to a cation, followed by loss of a proton, or, alternatively, by elimination of HCl from an intermediate chloride [of type **(74)**].

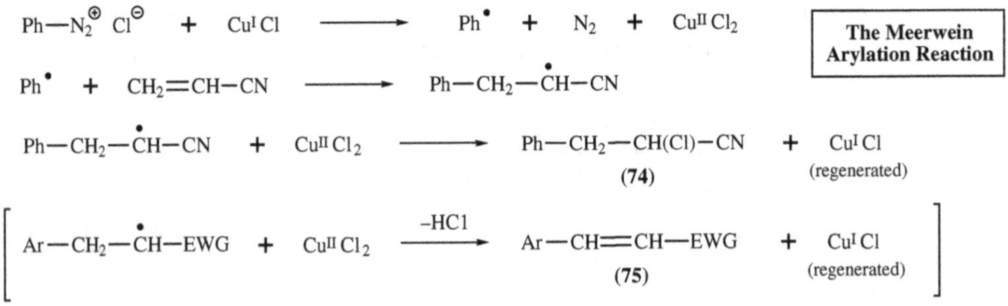

Fig. 8.4

Electron-rich double bonds, such as those present in enol ethers, can be used to trap electrophilic radicals (Fig. 8.5). Thus, reaction of tributyltin hydride with malonyl chloride (76) generates an electrophilic radical (adjacent to two electron-withdrawing ester substituents) and this can add, for example, to n-butyl vinyl ether to give (77) in 57% yield.

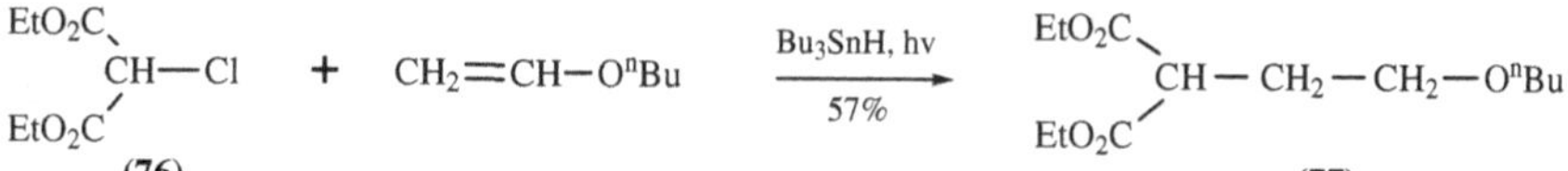

B. Giese, H. Horler and M. Leising, *Chem. Ber.*, 1986, **119**, 444–452

Fig. 8.5

Manganese(III) reagents are particularly good for preparing electrophilic radicals from carbonyl precursors. For example, reaction of manganese(III) acetate with 1,3-cyclohexanedione generates a radical at the α-carbon, which can add to an electron-rich alkene as shown in Fig. 8.6. A second equivalent of manganese(III) can oxidize the tertiary radical (78), and the resultant cation can be attacked by the enol oxygen atom to form the bi-cyclic product (79).

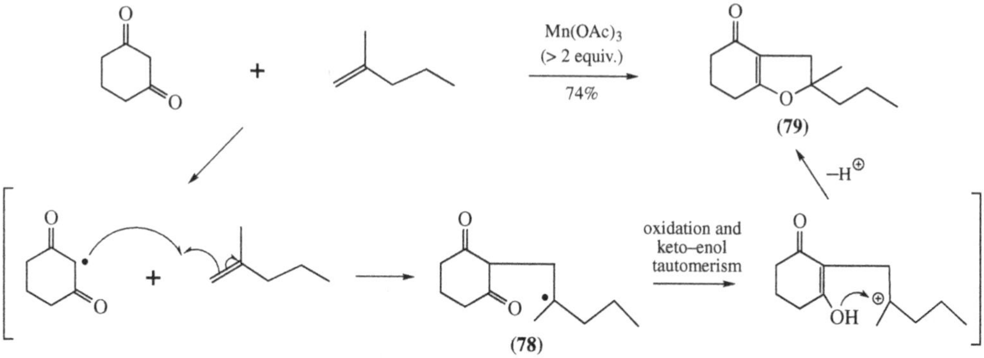

E.I. Heiba and R.M. Dessau, *J. Org. Chem.*, 1974, **39**, 3456–3457

Fig. 8.6

8.3 Stereoselectivity

As carbon-centred radicals are generally planar, reaction can take place from either face of the radical to give a mixture of two stereoisomers

(Fig. 8.7). Reaction of tributyltin hydride with a prochiral[1] radical centre often leads to fast hydrogen atom transfer and the formation of both enantiomers [(80) and (81)] in equal amounts. It can therefore be concluded that the transition states leading to the enantiomeric products are equal in energy. In some cases, however, the energies of the transition states are not equal; therefore reactions can be stereoselective (i.e. one enantiomer is formed in preference to the other) and attack from one face is preferred over the other.

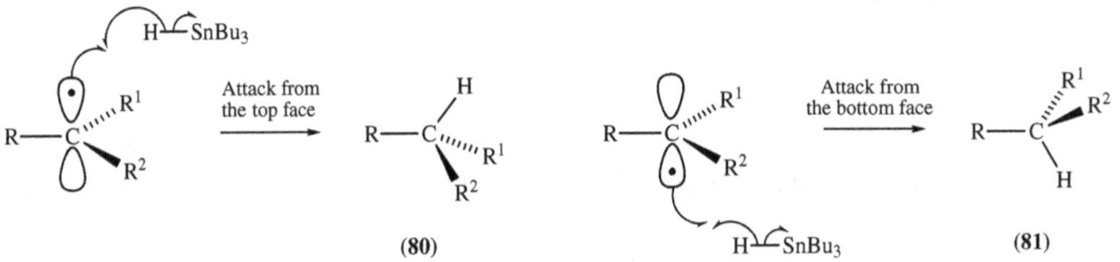

Fig. 8.7

8.3.1 Cyclic radicals

The two faces of the radical are different when the radical centre is adjacent to a chiral centre. For cyclic radicals, the chiral centre is 'fixed' with respect to the radical centre because the carbon–carbon bond (which joins them) cannot freely rotate. The radical will therefore react at the sterically least hindered face (in an *anti* attack) to avoid the largest substituent at the adjacent chiral centre. Thus, reaction of bromide (82) with allyltributyl-stannane leads to the *trans* product (84) because the alkene approaches radical (83) from the opposite side to the sulfonyl group (Fig. 8.8). In this allylation reaction, radical addition to the allyltin reagent produces a radical that can rapidly fragment to regenerate the tributyltin radical, which can then react with another molecule of bromide (82).

---

[1] A prochiral centre is the centre in an achiral precursor which, on reaction with a reagent, is converted to a chiral centre in the product.

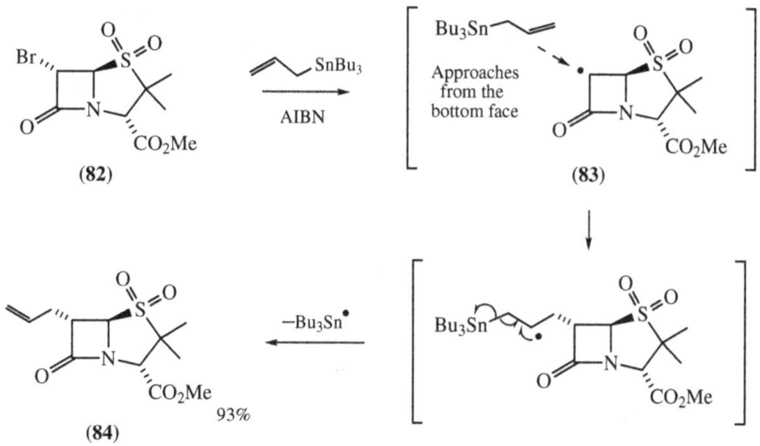

S. Hanessian and M. Alpegiani, *Tetrahedron*, 1989, **45**, 941–950

Fig. 8.8

Preferential addition to the less shielded face is also common for radicals centred on five- and six-membered rings. For cyclohexyl radicals, an equatorial β-substituent (on a carbon atom adjacent to the radical centre) promotes equatorial addition, whereas an axial β-substituent increases axial addition. For example, reaction of the thiocarbonyl derivative **(85)**, with an axial methyl substituent, leads to preferential addition of acrylonitrile to the bottom face of the ring (so as to avoid steric interactions with the methyl group) to give predominantly the axial product (Fig. 8.9).

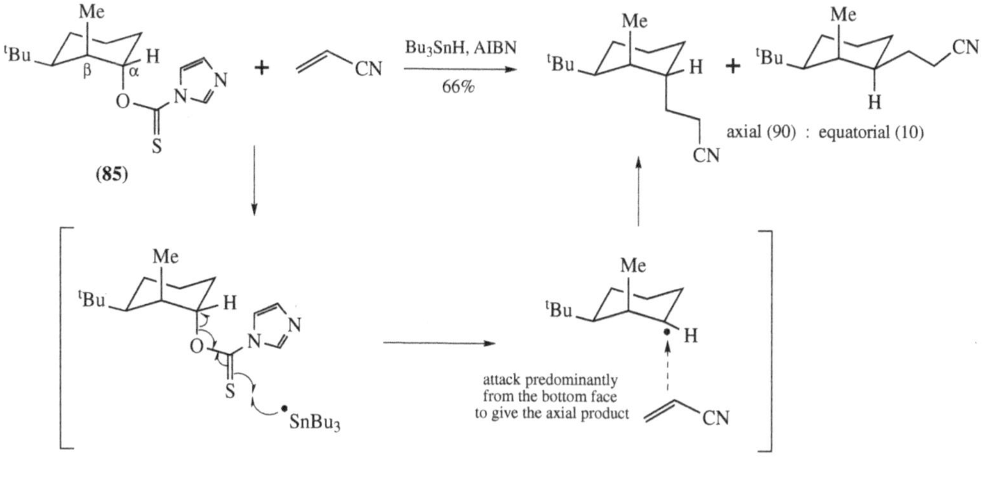

W. Damm, B. Giese, J. Hartung, T. Hasskerl, K.N. Houk, O. Hüter and H. Zipse, *J. Am. Chem. Soc.*, 1992, **114**, 4067–4079

Fig. 8.9

Stereoselectivity is not only governed by steric factors. For carbohydrates, for example, a stereoelectronic effect can explain the high levels of stereoselectivity that are observed in reactions at the anomeric centre. When the glucosyl bromide (86) reacts with tributyltin hydride, the radical at the anomeric centre is attacked predominantly from the bottom face to give the axial product. Hence, reaction with acrylonitrile leads to the α-C-glucoside (87) in 72% yield (Fig. 8.10). The corresponding β-C-glucoside (with the same carbon chain in the equatorial position) is only isolated in 5% yield. This is surprising because attack of the alkene (from the top face) to give the β-C-glucoside avoids 1,3-diaxial interactions, which might be expected to hinder the formation of the major product (87). However, the selectivity can be explained by interaction of the p orbital of the radical with the lone pair on the ring oxygen and the σ* orbital of the β-C—O bond (of the acetate group at the C-2 position). This (stabilizing) interaction is maximized when, on radical generation, the ring adopts the boat conformation (88) to allow the orbital of the radical and the σ* orbital to be in the same plane. Acrylonitrile then approaches radical (88) from the least hindered (bottom) face to give the C-1 alkylated product, which then 'ring flips' to give (87); the equatorial group in the boat conformation becomes an axial group in the chair conformation.

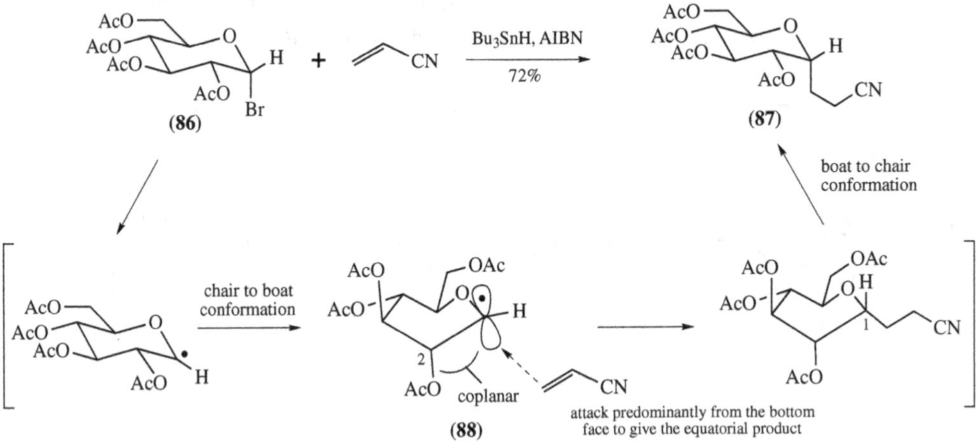

B. Giese and J. Dupuis, *Angew. Chem., Int. Ed. Engl.*, 1983, **22**, 622–623

Fig. 8.10

8.3.2 Acyclic radicals

The stereoselective reaction of acyclic radicals is much more difficult to achieve because the carbon–carbon bond connecting the radical and adjacent chiral centre can rotate (Fig. 8.11). Therefore, even though radical **(89)** reacts in an *anti* manner, and the tin hydride enters from the side opposite to the largest group (L), if the carbon–carbon bond can freely rotate, a 1 : 1 mixture of the diastereoisomers **(90)** and **(91)** will be formed.

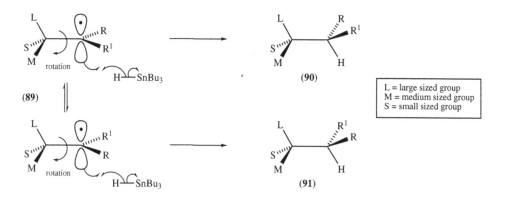

Fig. 8.11

Rotation around the carbon–carbon bond must therefore be slowed down or prevented, and this has been observed for substrates with bulky substituents at the radical centre and adjacent chiral centre. This is because steric interactions between these two groups favour one particular conformer, as illustrated by the stereoselective reduction of bromo-ester **(92)** using tributyltin hydride (Fig. 8.12). The intermediate radical **(93)** is biased towards a conformation in which the ester substituent is almost in the same plane as the hydrogen atom on the adjacent chiral centre. This is favoured because the steric interactions (known as allylic or $A^{1,3}$ strain) are less than for the alternative conformations, which have the large ester group closer to the bulkier alkyl [*tert*-butyl (^tBu) or methyl (Me)] substituents. Tributyltin hydride therefore approaches radical **(93)** from the less hindered face, to avoid the largest (^tBu) substituent, and hydrogen atom abstraction proceeds with excellent stereoselectivity.

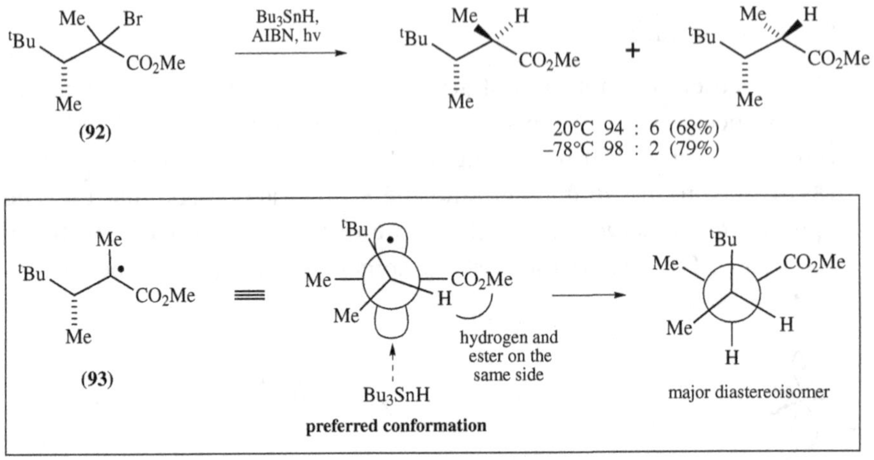

B. Giese, W. Damm, F. Wetterich and H–G. Zeitz, *Tetrahedron Lett.*, 1992, **33**, 1863–1866

Fig. 8.12

Reactions of acyclic radicals are carried out at low temperature to slow the rate of bond rotation and, in general, the lower the temperature, the better the stereoselectivity (see Section 5.2.5). For tributyltin hydride chemistry, the reaction between triethylborane (Et_3B) and oxygen is often employed as an initiator [rather than azobisisobutyronitrile (AIBN)/hν] at low temperatures (Fig. 8.13). The formation of a strong boron–oxygen bond (808 kJ mol^{-1}) provides the driving force for the formation of an ethyl radical which then reacts with tributyltin hydride.

$$Et_2B{-}Et \; + \; {}^{\bullet}O{-}O^{\bullet} \longrightarrow Et^{\bullet} \; + \; Et_2B{-}O{-}O^{\bullet}$$

$$Et^{\bullet} \; + \; {}^{\bullet}O{-}O^{\bullet} \longrightarrow Et{-}O{-}O^{\bullet}$$

$$Et{-}O{-}O^{\bullet} \; + \; Et_2B{-}Et \longrightarrow Et{-}O{-}O{-}BEt_2 \; + \; Et^{\bullet}$$

$$Bu_3Sn{-}H \; + \; Et^{\bullet} \longrightarrow Bu_3Sn^{\bullet} \; + \; Et{-}H$$

Fig. 8.13

The conformation of a radical can also be controlled by dipole–dipole repulsion, and this explains the stereoselective alkylation of iodide **(94)** (Fig. 8.14). This reaction uses a (removable) chiral auxiliary known as Oppolzer's sultam, and the intermediate radical adopts conformation **(95)** because the oxygen atom of the amide carbonyl and the oxygen atoms on the sulfur atom point in opposite directions (to minimize dipole repulsion). As a consequence, rotation around the amide bond is restricted. The methyl substituent at the radical centre prefers to avoid the bulky nitrogen substituent, and the allylstannane approaches from the less hindered top face; the bottom face is shielded by an axial oxygen of the SO_2 group. These features contribute to give an excellent selectivity.

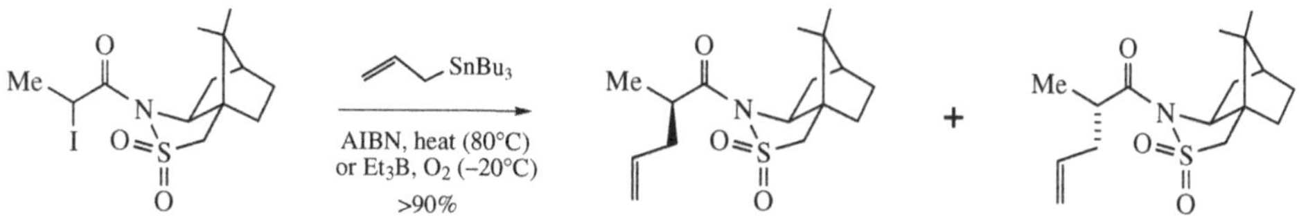

80°C 12 : 1
−20°C 25 : 1

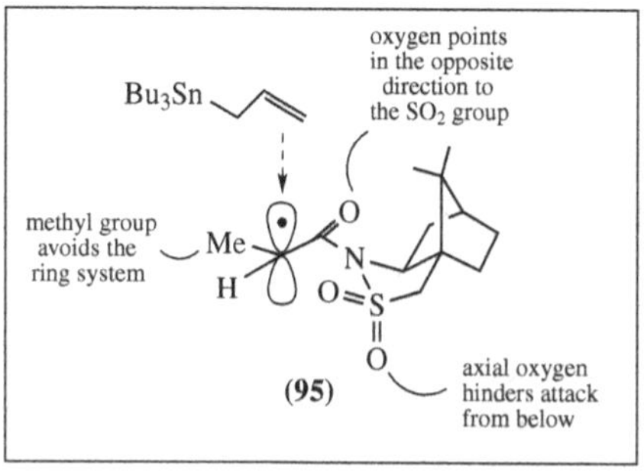

D.P. Curran, W. Shen, J. Zhang and T.A. Heffner, *J. Am. Chem. Soc.,* 1990, **112**, 6738–6740

Fig. 8.14

Hydrogen bonding and complexation can also restrict the rotation of bonds and, for example, a Lewis acid (such as magnesium bromide) can control the conformer population of oxazolidinone (96) by coordinating to the two carbonyl oxygens (Fig. 8.15). Conjugate addition of the ethyl radical to the double bond produces radical (97), which adopts a conformation in which the propyl chain is on the opposite face to the bulky nitrogen substituent. Addition of the allylstannane takes place from the top face so as to avoid the large aromatic groups (on the oxazolidinone ring) to give (98) in excellent yield.

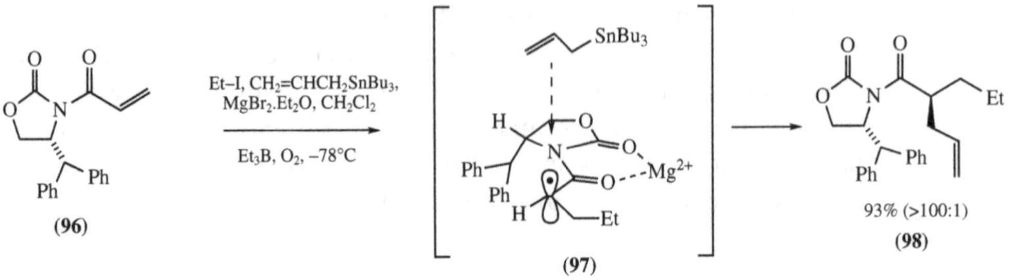

M.P. Sibi and J. Ji, *J. Org. Chem.*, 1996, **61**, 6090–6091

Fig. 8.15

8.4 Synthetic applications

Some examples of intermolecular additions will now be discussed in more detail to show the importance of these types of reaction in synthesis.

(–)-*exo*-Brevicomin is a naturally occurring pheromone produced by a North American bark beetle. A short approach to the (+)-enantiomer **(102)**, from inexpensive (*R,R*)-tartaric acid **(99)**, has made use of a radical addition reaction in the key step (Fig. 8.16). The acid **(99)** is first converted to primary iodide **(100)** and then reacted with methyl vinyl ketone in the presence of tributyltin hydride and AIBN. The tin hydride is added slowly (over 1.5 h) to minimize simple reduction and an excess of the electron-poor alkene is used to increase the rate of addition of the nucleophilic radical. Under these conditions, regioselective 1,4-addition of the radical to the double bond gives **(101)** in reasonable yield. The ketone group in **(101)** is subsequently used to form the acetal functionality required for **(102)**.

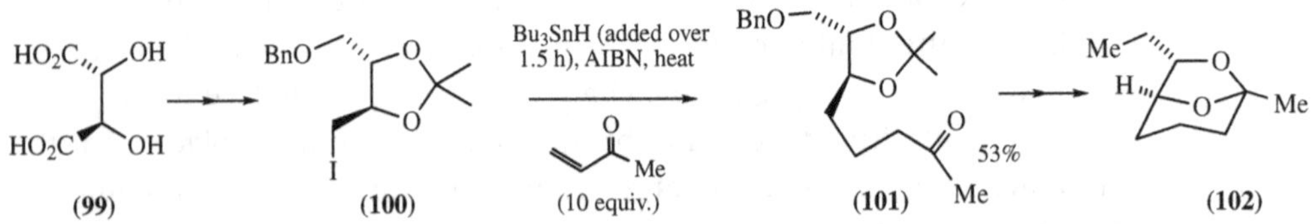

B. Giese and R. Rupaner, *Synthesis*, 1988, 219–221

Fig. 8.16

Intermolecular addition can be combined with intramolecular cyclization reactions to provide an efficient 'one-pot' approach to complex targets.

This has been used to prepare camptothecin (103), an alkaloid which exhibits anticancer properties (Fig. 8.17). Photolysis of hexamethylditin with aryl bromide (104) generates radical (105), and this adds intermolecularly to the carbon atom of phenyl isonitrile (which is present in excess). Radical (105) does not undergo an intramolecular cyclization (onto the alkyne triple bond) as this would form a very strained four-membered ring. However, the resulting imidoyl radical (106) is able to cyclize onto the triple bond, as this forms a more stable five-membered ring, to give the vinyl radical (107). Finally, cyclization onto the benzene ring produces the cyclohexadienyl radical (108), which can then be oxidized to give (109). This impressive sequence (intermolecular reaction followed by intramolecular reaction) assembles the natural product ring system to give the tetracycle (109) in 45% yield from (104).

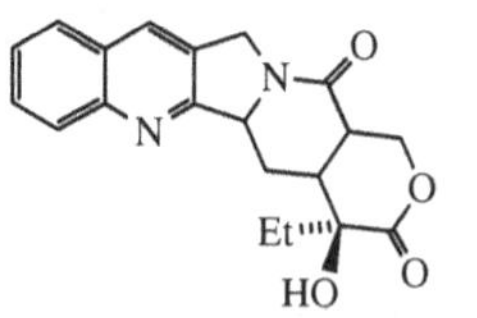

Camptothecin (103)

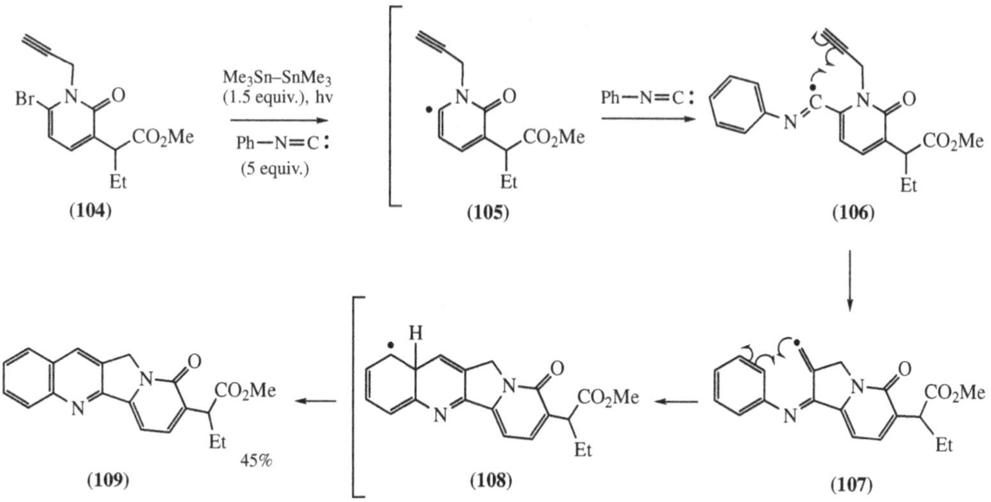

D.P. Curran and H. Liu, *J. Am. Chem. Soc.*, 1992, **114**, 5863–5864

Fig. 8.17

A related strategy has been used to access the prostaglandin family of natural products. These compounds control a variety of physiological responses (including blood clotting and muscle contraction) in the human body. In this approach, (+)-prostaglandin $F_{2\alpha}$ **(110)** is prepared by an intramolecular cyclization, followed by an intermolecular addition to an unsaturated ketone (Fig. 8.18). Heating iodide **(111)** with tributyltin hydride (derived from reduction of the chloride with sodium cyanoborohydride) produces primary radical **(113)** which cyclizes onto the alkene to form the bi-cyclic radical **(114)**. A low concentration of tributyltin hydride, generated from only 0.1 equivalent of the chloride, is required to minimize (simple) reduction of **(113)** and, particularly, reduction of **(114)**. The cyclization produces a *cis* ring junction, which minimizes ring strain, and radical **(114)** is trapped by reaction with octenone **(112)**. Even though octenone **(112)** is present in excess, this does not lead to significant trapping of radical **(113)** because the intramolecular cyclization [to form **(114)**] is much faster than the competing intermolecular addition. Therefore, the octenone approaches radical **(114)** from the less hindered top face, so as to avoid the tetrahydrofuran ring and the bulky protected oxygen substituent. Finally, reaction between **(115)** and tributyltin hydride completes the synthesis of the prostaglandin precursor **(116)**. This approach has introduced two side chains on the cyclopentane ring with the correct (prostaglandin) stereochemistry in a single step. In addition, the trimethylsilyl substituent in **(116)** can be transformed to an alkene and the ketone can be stereoselectively reduced to give the desired C-12 side chain of the natural product. The C-8 side chain can also be introduced by manipulation of the cyclic acetal group in **(116)**.

The formation of prostaglandins in nature is also believed to involve radical intermediates, and this biochemical pathway is known as the 'arachidonic acid cascade' (Fig. 8.19). The unsaturated fatty acid, arachidonic acid **(117)**, is thought to react selectively with molecular oxygen, in the presence of enzymes, to form a very reactive hydroperoxide **(119)**

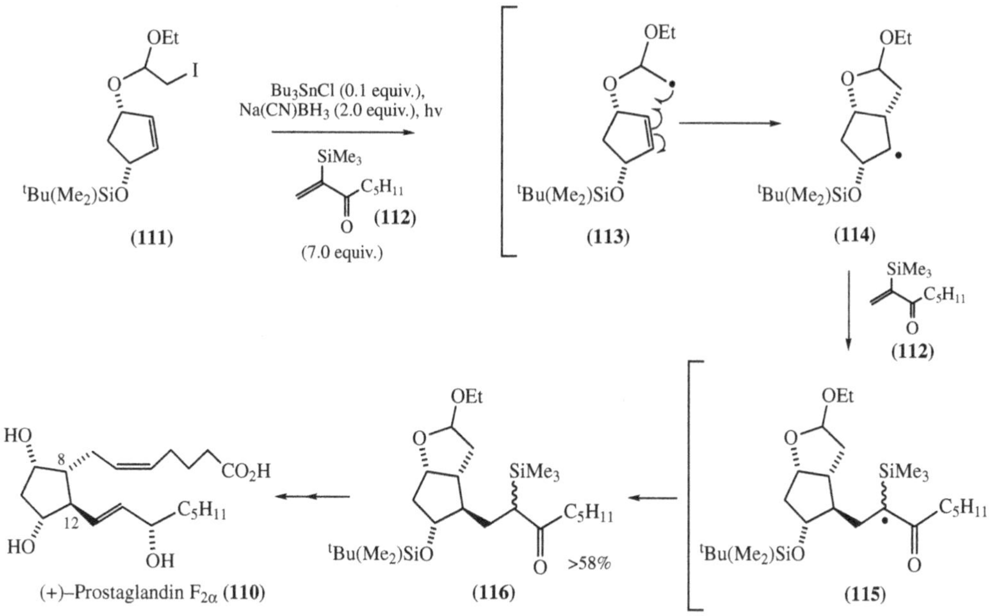

G. Stork, P.M. Sher and H.-L. Chen, *J. Am. Chem. Soc.*, 1986, **108**, 6384–6385

Fig. 8.18

(which has been isolated). This process involves a selective hydrogen atom abstraction (at C-13) to give a resonance-stabilized radical that can combine with oxygen. The resulting oxygen-centred radical **(118)** can then undergo a tandem cyclization reaction to form two five-membered rings stereoselectively. This cyclization is followed by addition of the (allylic) radical to a further molecule of oxygen to give, after hydrogen atom abstraction, the hydroperoxide **(119)**. The hydroperoxide can react further to give prostaglandins, including prostaglandin $F_{2\alpha}$ ($PGF_{2\alpha}$) **(110)**.

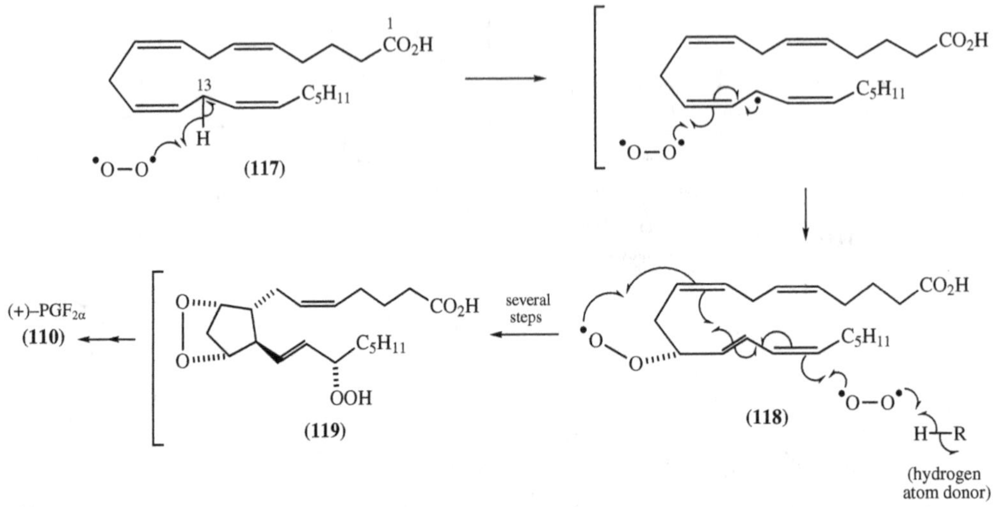

Fig. 8.19

The most important industrial application of radical chemistry is in polymerization. This exploits the intermolecular addition of radicals to alkenes, such as styrene or methyl methacrylate, to form high molecular weight polymers (Fig. 8.20). A peroxide or azo compound is generally used to initiate the polymerization and, after a number of propagation steps, termination reactions stop the growing chain. The terminations can take place at any time during the polymerization to give polymers with a variety of chain lengths. Therefore, these conventional polymerizations do not allow the synthesis of well-defined polymers (having chains of the same length) with controlled architectures that could exhibit advantageous physical properties.

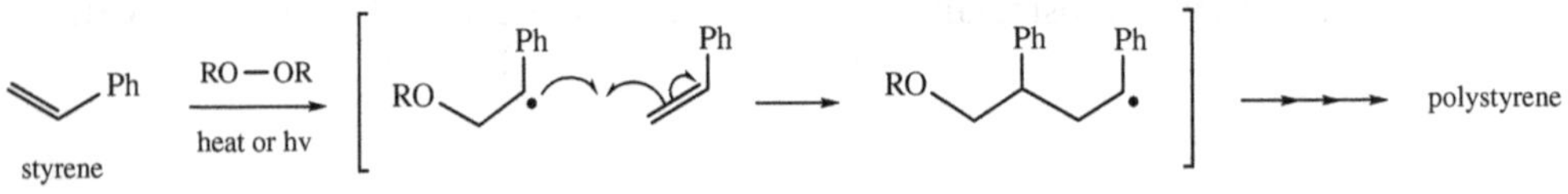

Fig. 8.20

One approach aimed at controlling the polymerization process, by generating only a low concentration of radicals, is based on a 'living' radical polymerization reaction (Fig. 8.21). An atom transfer reaction between a copper(I) complex and an alkyl chloride can be used to initiate the polymerization of styrene. This is a fast and reversible reaction [step (i)] that ensures a low concentration of the alkyl radical **(120)** which, in turn, minimizes radical termination (e.g. coupling or disproportionation) reactions. Therefore, alkyl radical **(120)** reacts very efficiently with the styrene double bond to give a benzylic radical **(121)** [step (ii)], and this is then trapped by halogen atom abstraction from the copper(II) complex [step (iii)]. This rapid interchange, to give secondary chloride **(122)**, occurs before any further addition to styrene can take place. Further chlorine atom abstraction followed by radical addition [step (iv)] to give **(123)**, and trapping, allows another molecule of styrene to be incorporated into the growing polymer to give **(124)** [step (v)]. As the chain always contains a chlorine atom at the end, this process can be repeated further [step (vi)], and the polymer can be thought of as 'living' (until all the styrene has reacted).

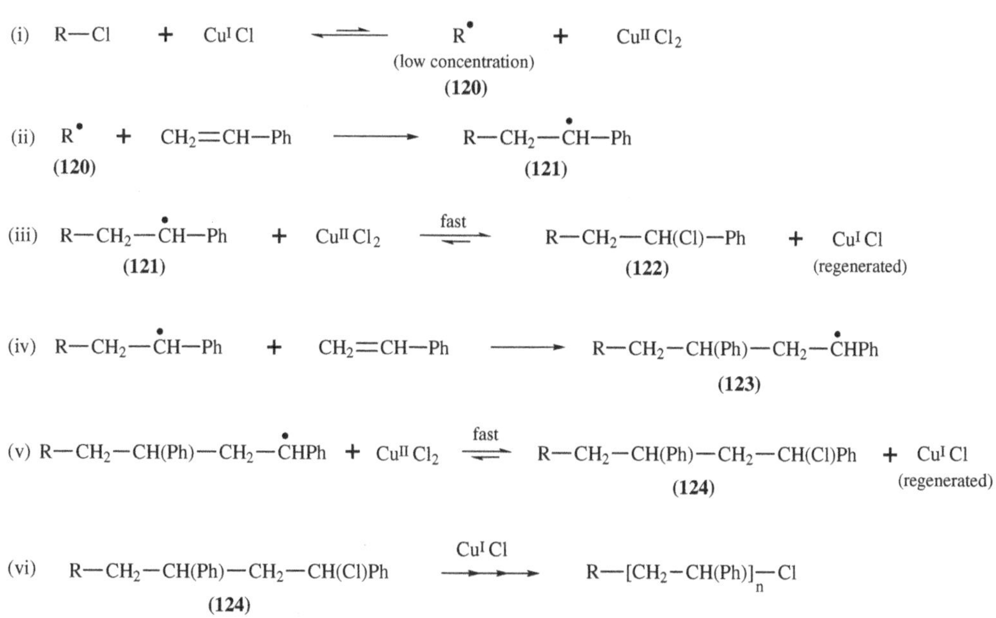

(i) $R-Cl$ + $Cu^I Cl$ $\rightleftharpoons$ $R^\bullet$ + $Cu^{II} Cl_2$
(low concentration)
(120)

(ii) $R^\bullet$ + $CH_2{=}CH-Ph$ $\longrightarrow$ $R-CH_2-\overset{\bullet}{C}H-Ph$
(120)   **(121)**

(iii) $R-CH_2-\overset{\bullet}{C}H-Ph$ + $Cu^{II} Cl_2$ $\overset{fast}{\rightleftharpoons}$ $R-CH_2-CH(Cl)-Ph$ + $Cu^I Cl$
(121)   **(122)**   (regenerated)

(iv) $R-CH_2-\overset{\bullet}{C}H-Ph$ + $CH_2{=}CH-Ph$ $\longrightarrow$ $R-CH_2-CH(Ph)-CH_2-\overset{\bullet}{C}HPh$
(123)

(v) $R-CH_2-CH(Ph)-CH_2-\overset{\bullet}{C}HPh$ + $Cu^{II} Cl_2$ $\overset{fast}{\rightleftharpoons}$ $R-CH_2-CH(Ph)-CH_2-CH(Cl)Ph$ + $Cu^I Cl$
(124)   (regenerated)

(vi) $R-CH_2-CH(Ph)-CH_2-CH(Cl)Ph$ $\overset{Cu^I Cl}{\longrightarrow}$ $R-[CH_2-CH(Ph)]_n-Cl$
(124)

K. Matyjaszewski, T.E. Patten and J. Xia, *J. Am. Chem. Soc.*, 1997, **119**, 674–680

Fig. 8.21

8.5 Summary

Although intermolecular additions are more difficult to conduct than cyclization reactions, many examples of efficient carbon–carbon bond-forming reactions are known. These reactions rely on matching the polarity of the radical to the acceptor double (or triple) bond; a nucleophilic radical needs to react with an electron-poor double bond and vice versa.

The experimental conditions are also very important, and an excess of the alkene or alkyne is usually required in order to increase the rate of addition and minimize competitive atom abstraction reactions. Synthetically useful tandem sequences can also be employed, and these usually involve a fast intramolecular cyclization followed by a slower intermolecular addition reaction.

Further reading

Giese, B. (1985) Syntheses with radicals—C—C bond formation via organotin and organomercury compounds. *Angewandte Chemie, International Edition in English*, **24**, 553–565.

Giese, B., Damm, W. & Batra, R. (1994) Allylic strain effects in stereoselective radical reactions. *Chemtracts-Organic Chemistry*, **7**, 355–370.

Hawker, C.J. (1997) 'Living' free radical polymerization: a unique technique for the preparation of controlled macromolecular architectures. *Accounts of Chemical Research*, **30**, 373–382.

Porter, N.A., Giese, B. & Curran, D.P. (1991) Acyclic stereochemical control in free-radical reactions. *Accounts of Chemical Research*, **24**, 296–304.

CHAPTER 9

Radical Translocation Reactions

9.1 Introduction

Radical translocation or intramolecular abstraction reactions involve the movement of a radical centre from its site of generation to another atom, usually located five or six atoms away. Radical translocation can proceed, for example, by abstraction of a hydrogen atom from a carbon–hydrogen bond in an $S_{H}i$ reaction (substitution, homolytic, intramolecular) (Fig. 9.1). A near-linear attack maximizes the interaction of the radical and C—H bond orbitals, and this favours the formation of six- and seven-membered transition states leading to 1,5- and 1,6-hydrogen atom transfers, respectively. The reaction is energetically favoured by the formation of a more stable radical. For example, a reactive vinyl or aryl radical ($R^{1\bullet}$) will readily abstract a hydrogen atom to form an alkyl radical ($R^{2\bullet}$), and this provides a route to (alkyl) radicals that might otherwise be difficult to prepare directly.

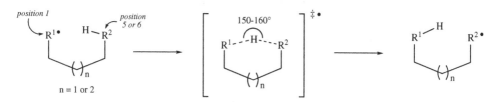

Fig. 9.1

9.2 Methods for radical translocation

9.2.1 Tin hydride

Reaction of aryl or vinyl halides with tributyltin hydride produces reactive radicals that can undergo 1,5- or 1,6-hydrogen atom transfer. The translocation leads to the formation of more stable primary, secondary or, preferably, tertiary alkyl radicals; the more stable the alkyl radical, the greater the driving force for translocation. For successful translocation, the aryl or vinyl radical must abstract the hydrogen atom before reaction with tin hydride (leading to simple reduction) can take place. This is usually not a major problem because intramolecular translocation is often much faster than intermolecular reduction using tin hydride. In addition, the rate of abstraction can be increased if the radical is held in close proximity to the reacting carbon–hydrogen bond.

This method can be illustrated by the reaction of vinyl bromide (125) with tributyltin hydride and azobisisobutyronitrile (AIBN) (Fig. 9.2). Abstraction of the bromine atom by the tributyltin radical leads to vinyl radical (126). This intermediate vinyl radical can be reduced to give alkene (127), but the use of a low concentration of tributyltin hydride (generated from only 0.1 equivalent of the chloride) ensures that this undesired product is only formed in 23% yield. Instead, the vinyl radical (126) preferentially undergoes a regioselective 1,5-hydrogen atom transfer to form the secondary radical (128), which is stabilized by the ester substituent. This transfer step has a rate constant of around 10^6 s^{-1} (at 80°C). Reaction of (128) with tributyltin hydride could also lead to alkene (127), but a faster cyclization reaction can take place to form the cyclopentane radical (129). The presence of a geminal diester (i.e. two ester groups on the same carbon) can help both the atom abstraction and cyclization reactions by restricting the conformation of the carbon chain (Thorpe–Ingold effect). This ensures that the vinyl radical in (126) and the alkyl radical in (128) are held in close proximity to the carbon–hydrogen and alkene bonds, respectively. Finally, reduction of the primary radical (129) with tributyltin hydride affords the substituted cyclopentane (130) in 60% yield as predominantly the cis diastereoisomer. The cis stereochemistry is predicted from a chair-like transition state (see Section 7.4.1).

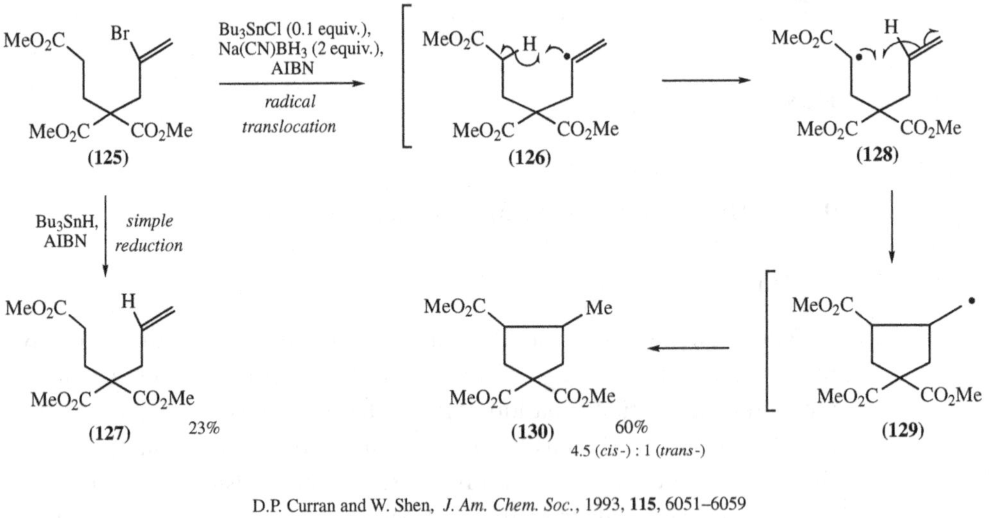

D.P. Curran and W. Shen, J. Am. Chem. Soc., 1993, 115, 6051–6059

Fig. 9.2

In summary, although tin radicals cannot abstract hydrogen atoms directly from carbon–hydrogen bonds, this method of translocation has *indirectly* allowed the selective formation of a radical from a carbon–hydrogen bond.

A similar radical translocation reaction has been used to prepare the bi-cyclic ring system (131), which is found in the pyrrolizidine alkaloid family of natural products (Fig. 9.3). Radical generation from iodide (132) generates vinyl radical (133) which can translocate, by a 1,6-hydrogen atom transfer, to give the more stable allylic radical (134). This radical is stabilized by delocalization of the electron over the alkene double bond and ester carbonyl. Radical (134) can attack the alkene to form a five-membered ring leading to the tri-cycle (135) in 60–85% yield. Subsequent oxidative cleavage of the double bond in (135) (followed by reduction using sodium borohydride) gives the desired bi-cyclic ring system (131).

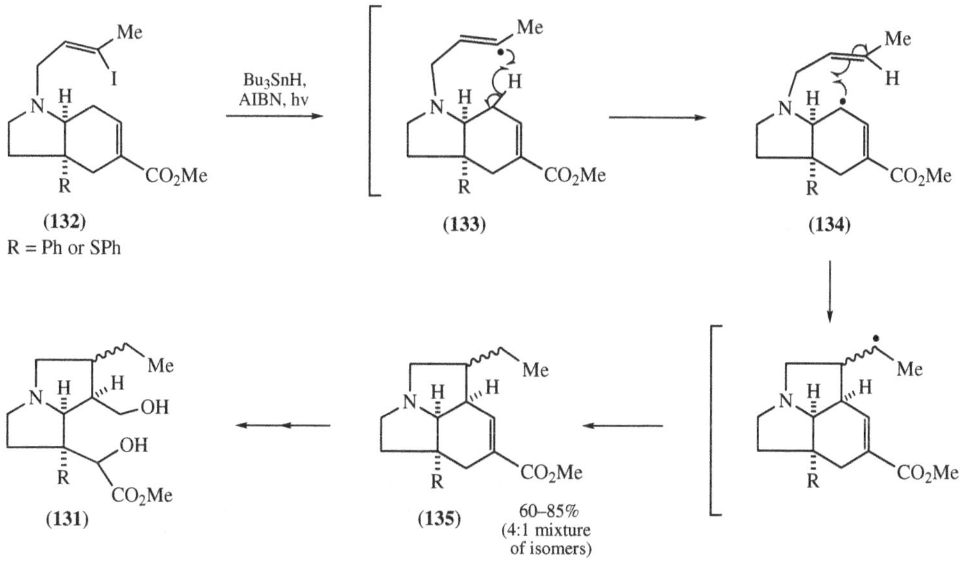

D.C. Lathbury, P.J. Parsons and I. Pinto, *J. Chem. Soc., Chem. Commun.*, 1988, 81–82

Fig. 9.3

Reactive aryl radicals can undergo similar translocation reactions, and a particularly elegant application of this method involves the use of 'protecting/radical translocating' (PRT) groups.

A number of common oxygen and nitrogen protecting groups, such as benzyl and benzoyl groups, contain an aromatic ring. These groups are frequently used in synthesis because they are readily introduced into a molecule, they are resistant to a variety of reaction conditions and efficient methods are available for their removal. If a halogen atom is incorporated in the benzene ring, the protecting group can also function as the radical precursor. This strategy has been utilized to alkylate cyclic amines, such as pyrrolidine **(136)**, at the α-position (Fig. 9.4).

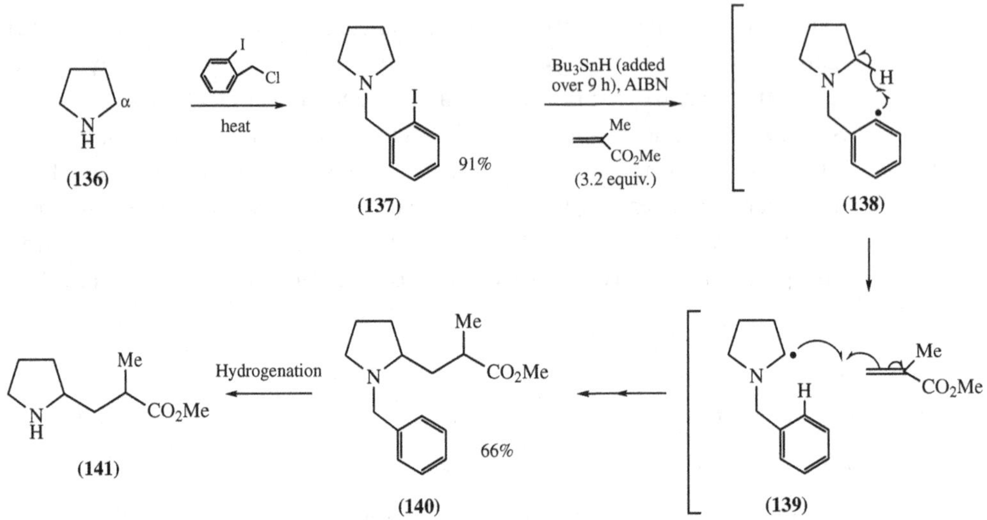

L. Williams, S.E. Booth and K. Undheim, *Tetrahedron*, 1994, **50**, 13697–13708

Fig. 9.4

The benzyl protecting group is first introduced by an alkylation reaction to give the 2-iodobenzyl precursor **(137)**. Subsequent reaction with tributyltin hydride generates aryl radical **(138)** and this undergoes a 1,5-hydrogen atom transfer to give cyclic radical **(139)**, which is a nucleophilic radical stabilized by the adjacent nitrogen atom. Intermolecular addition to an electron-poor alkene, such as methyl methacrylate, can successfully compete with hydrogen atom abstraction to give the α-alkylated pyrrolidine **(140)** in 66% yield. During the process, the N-iodobenzyl group in **(137)** has been converted to an N-benzyl group in **(140)**, and this can subsequently be removed using, for example, hydrogenation to give the α-alkylated pyrrolidine **(141)**.

Similar transformations can be carried out using 2 equivalents of samarium(II) iodide (Fig. 9.5). The first equivalent generates aryl radical **(142)** which undergoes translocation, while a second equivalent results in an electron transfer reaction to give the α-amino-organosamarium(III) compound **(143)**. These organosamarium compounds behave as carbanions and therefore, for example, they will add to ketones to give tertiary alcohols after acidic work-up.

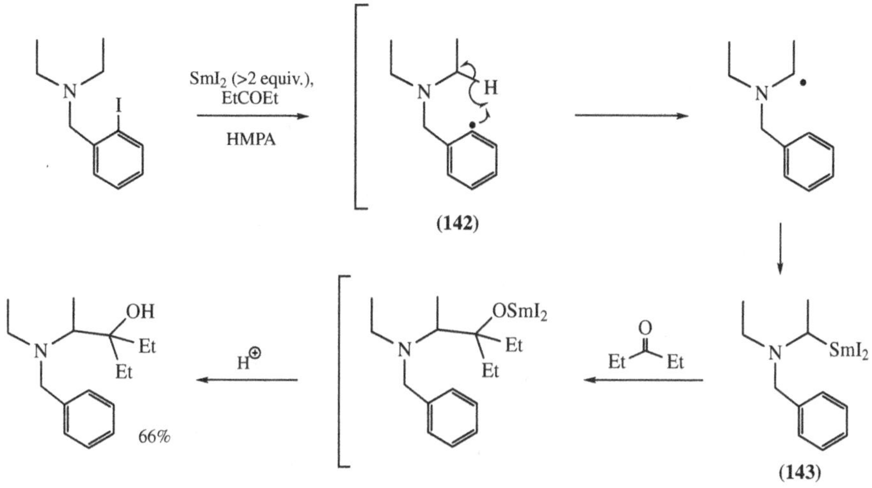

M. Murakami, M. Hayashi and Y. Ito, *J. Org. Chem.*, 1992, **57**, 793–794

Fig. 9.5

9.2.2 Photolysis of carbonyls

When a ketone is irradiated with visible or ultraviolet (UV) light, the resulting singlet or triplet excited state can undergo hydrogen atom abstraction reactions. Abstractions may take place intramolecularly from a C—H group spatially close to the excited carbonyl in the Norrish type II reaction.[1] For example, UV irradiation of 2-pentanone **(144)** produces an excited singlet or triplet diradical **(145)**, which undergoes a selective 1,5-hydrogen atom transfer to generate 1,4-diradical **(146)** (Fig. 9.6). The reaction proceeds via a six-membered transition state, and leads to the formation of a strong O—H bond together with a primary radical; similar abstractions to give more stable secondary and tertiary radicals are favourable. There are two pathways by which **(146)** can react so as to 'pair' the electrons: (i) by ring closure to form a strained four-membered cyclobutane ring (path a); or (ii) by fragmentation to form an alkene and

---

[1] The Norrish type I reaction leads to homolysis of the bond adjacent to the carbonyl group and is more commonly observed in the gas phase. For aliphatic ketones [RC(=O)R], this produces an alkyl radical (R•) together with an acyl radical [R—C(=O)•]. Products can arise from decarbonylation of the acyl radical, and dimerization and/or disproportionation of the alkyl radical.

an enol, which tautomerizes to the ketone (path b). The fragmentation step is analogous to the McLafferty rearrangement of carbonyl compounds, which is observed in mass spectrometry (see Section 11.2.4).

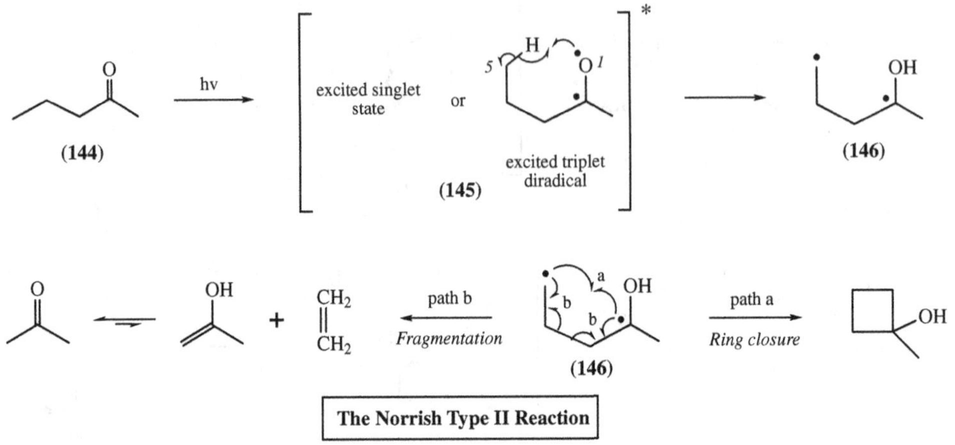

The Norrish Type II Reaction

Fig. 9.6

Steric factors play a large part in determining the ratio of fragmentation to ring closure. For efficient fragmentation to occur, the diradical must adopt a conformation in which the two singly occupied molecular orbitals (SOMOs) are (aligned) parallel to the orbital of the C_α—C_β bond that is broken (Fig. 9.7). Any steric factors that inhibit this conformation (and orbital alignment) will promote cyclization.

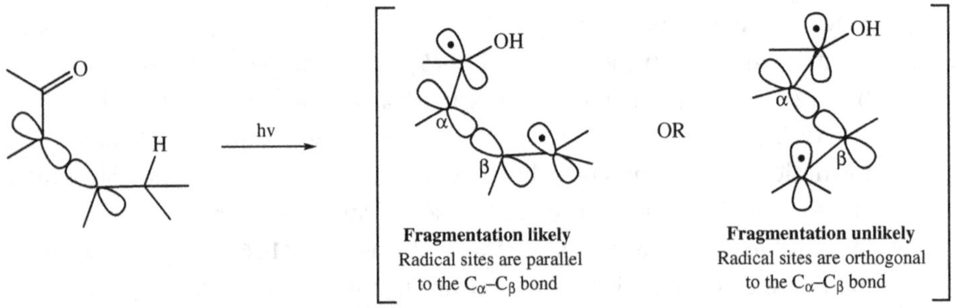

Fig. 9.7

The ring closure reaction is more important than fragmentation in synthesis and is widely used to form four-membered rings that are difficult to prepare using other methods. This is illustrated by the formation of the highly strained bi-cyclic diastereoisomers **(148)** and **(149)** from the photolysis of aryl ketone **(147)** (Fig. 9.8).

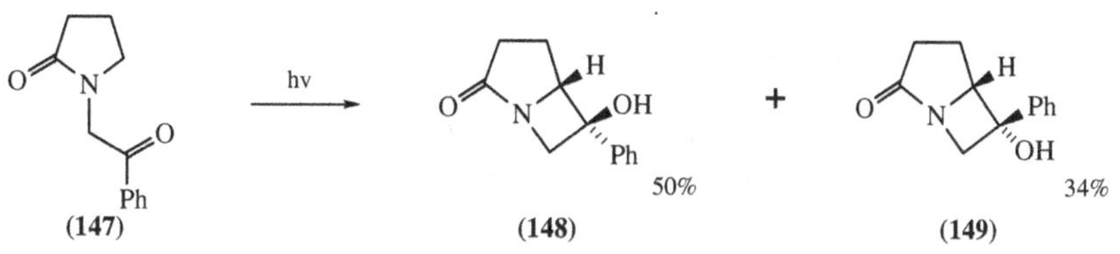

50%   34%

(147)   (148)   (149)

L. Ouazzani-Chadi, J-C. Quirion, Y. Troin and J-C. Gramain, *Tetrahedron*, 1990, **46**, 7751–7762

Fig. 9.8

When the carbonyl compound does not have any hydrogen atoms at the 5-position, abstraction of more remote hydrogen atoms is possible, and a 1,6-hydrogen atom transfer has been used to prepare (±)-paulownin **(150)** (Fig. 9.9). One of the heterocyclic rings of this lignan natural product is formed by a stereoselective photocyclization reaction of cyclic ketone **(151)**. The presence of an oxygen atom in the side chain of **(151)** prevents a (more favourable) 1,5-abstraction, and a benzylic hydrogen atom is abstracted to give diradical **(152)**. This diradical cannot undergo fragmentation as the radical centres are too far apart, and so cyclization proceeds to form the bi-cyclic ring in 64% yield. Only one stereoisomer is formed as cyclization (onto a rigid five-membered ring) results in a less strained *cis* fused ring system, with the aromatic substituents on the sterically less hindered top face.

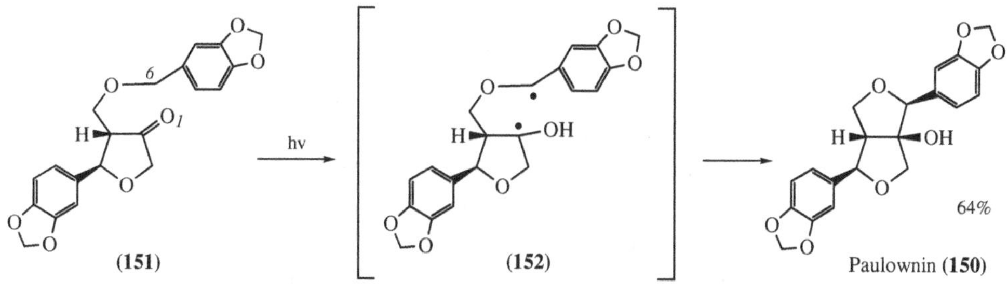

64%

(151)   (152)   Paulownin **(150)**

G.A. Kraus and L. Chen, *J. Am. Chem. Soc.*, 1990, **112**, 3464–3466

Fig. 9.9

9.2.3 Photolysis of nitrites

Irradiation of nitrite esters, such as **(153)**, leads to (reversible) homolysis of the weak nitrogen–oxygen bond to give an oxygen-centred radical **(154)** together with nitric oxide (Fig. 9.10). If this reactive alkoxyl radical is in close proximity to a hydrogen atom at position 5, a 1,5-hydrogen atom abstraction can take place to form a more stable carbon-centred radical **(155)**. The nitric oxide (formed on homolysis) can then combine with **(155)** to form nitroso alcohol **(156)**, which tautomerizes to oxime **(157)**. Oximes are convenient precursors to aldehydes, and acid hydrolysis of **(157)** produces hydroxyaldehyde **(158)**, which exists predominantly in the cyclic hemiacetal (or lactol) form. The unactivated methyl group in **(153)** is therefore selectively oxidized to aldehyde **(158)** under very mild conditions in a process known as the Barton reaction.

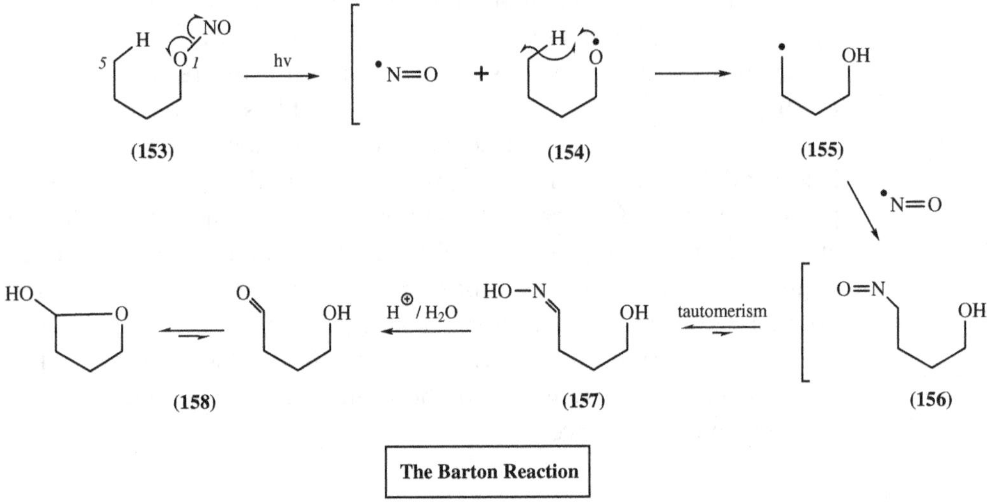

Fig. 9.10

This chemistry has been exploited in the steroid field and, for example, the nitrite ester of corticosterone acetate **(159)** can be selectively oxidized to aldosterone 21-acetate **(162)** (Fig. 9.11). The oxidation arises from the close proximity of the C-18 methyl substituent and oxygen-centred radical in **(160)**; these groups are held in a 1,3-diaxial relationship by the rigid steroid skeleton. Competitive oxidation of the C-19 methyl group, which is also close to the oxygen-centred radical, decreases the efficiency of the reaction to give **(161)** in 21% yield, and this can subsequently be hydrolysed to **(162)**.

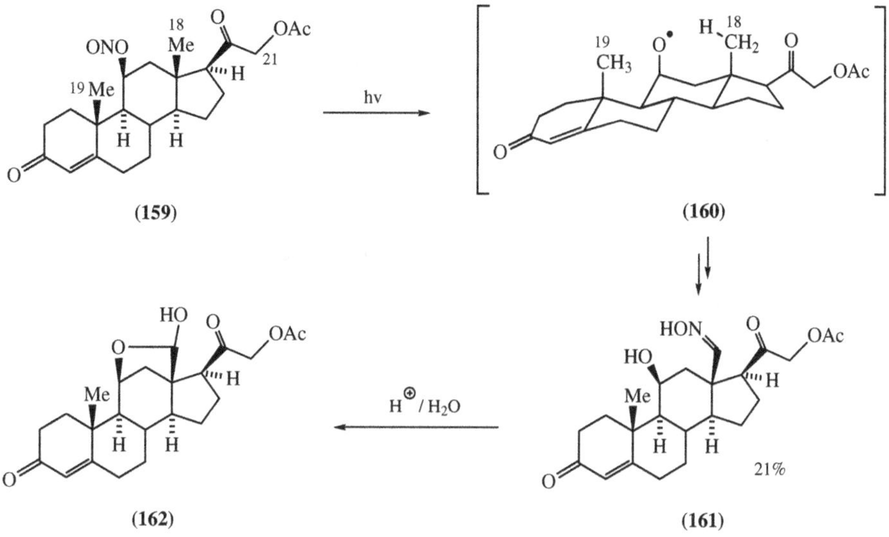

D.H.R. Barton and J.M. Beaton, *J. Am. Chem. Soc.*, 1960, **82**, 2641 and 1961, **83**, 4083–4089

Fig. 9.11

This 'remote functionalization' of an axial methyl group to form an aldehyde, regioselectively, mimics biochemical processes in nature. These enzyme-controlled reactions are able to selectively functionalize hydrocarbon segments of a molecule remote from any functional group.

9.2.4 Chlorination

The remote functionalization of rigid steroid molecules has been extended to chlorinations, using aryliodine chloride precursors. A reactive chlorine radical usually reacts unselectively with alkane C—H bonds to give alkyl radicals, which react further to give mixtures of alkyl chlorides. If, however, the geometry of attack of the chlorine atom can be controlled, by attachment to a rigid template, the chlorine atom may selectively abstract one particular hydrogen atom (Fig. 9.12).

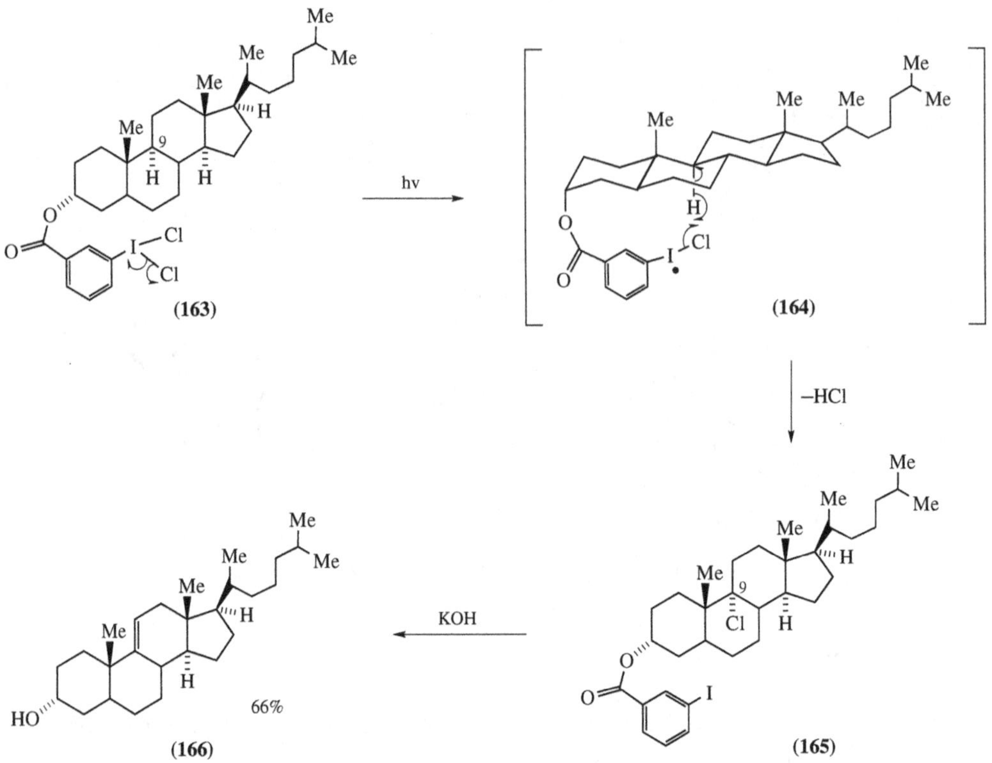

R. Breslow, *Acc. Chem. Res.*, 1980, **13**, 170–177

Fig. 9.12

Thus, irradiation of the dichloroiodo-ester derivative of α-cholestanol **(163)** generates radical **(164)**, in which a chlorine atom is associated with the iodine atom on the benzene ring. The iodo-ester group acts as a template to fix the position of the chlorine atom so that it is directly under the C-9 hydrogen atom of the steroid. The hydrogen atom is therefore selectively abstracted and the resulting tertiary radical then abstracts a chlorine atom from **(163)** (in a chain reaction) to give chloride **(165)**. Only one stereoisomer is produced as the rigid steroid skeleton ensures that the chlorine atom is introduced from the less hindered bottom face. Treatment of **(165)** with hydroxide promotes both an elimination reaction and ester hydrolysis to give 9(11)-cholestenol **(166)** in 66% yield [from **(163)**]. This work shows that even a reactive chlorine atom can selectively abstract a hydrogen atom, when attached to a template, in an intramolecular reaction.

9.3 Summary

Radical translocation reactions, particularly 1,5-hydrogen atom abstraction reactions, are very useful for the selective formation of radicals even in complex precursors. These translocations can lead to the introduction of functional groups at specific sites within a molecule. For a radical centre to be translocated, a more stable radical must be produced, and hydrogen transfer can often take place, for example, from carbon to carbon or carbon to oxygen. Reactions involving oxygen are strongly exothermic because a much stronger O—H bond is formed at the expense of a weaker C—H bond, and similar reactions are known involving hydrogen atom transfer from carbon to nitrogen (see Section 11.2.4).

Intramolecular 1,5- and 1,6-hydrogen atom abstractions are favoured because the six- or seven-membered transition state can be readily formed. In contrast, larger range hydrogen atom abstractions are only observed when the geometry of the radical is constrained, and this has found application in, for example, the remote functionalization of rigid steroid molecules.

Further reading

Curran, D.P., Yu, H. & Liu, H. (1994) Amide-based protecting/radical translocating (PRT) groups. Generation of radicals adjacent to carbonyls by 1,5-hydrogen transfer reactions of o-iodoanilides. *Tetrahedron*, **50**, 7343–7366.

Keukeleire, D.D. & He, S.-L. (1993) Photochemical strategies for the construction of polycyclic molecules. *Chemical Reviews*, **93**, 359–380.

Radical Anions

10.1 Introduction

The addition of an electron to a neutral (non-radical) molecule generates a species with both a negative charge and an unpaired electron. This is known as a radical anion. For the addition to take place, molecules must have a low-energy lowest unoccupied molecular orbital (LUMO) so that the electron can be readily accepted (Fig. 10.1). Typical precursors include aromatics, alkenes or carbonyl compounds, and the electron enters the π^* orbital (or LUMO) of, for example, an alkene or ketone double bond to give radical anions which can be represented by two resonance canonicals (Fig. 10.2).

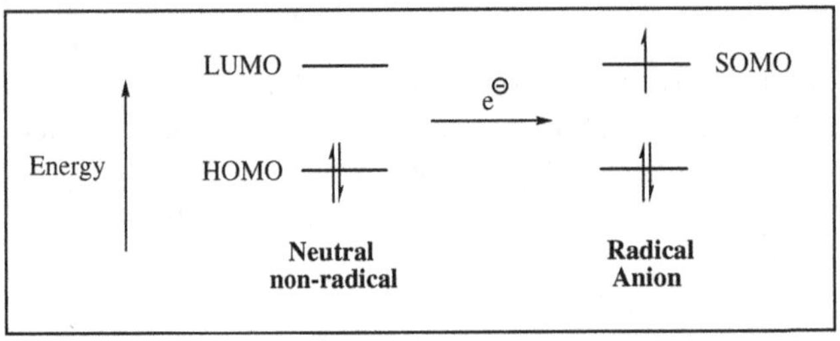

Fig. 10.1

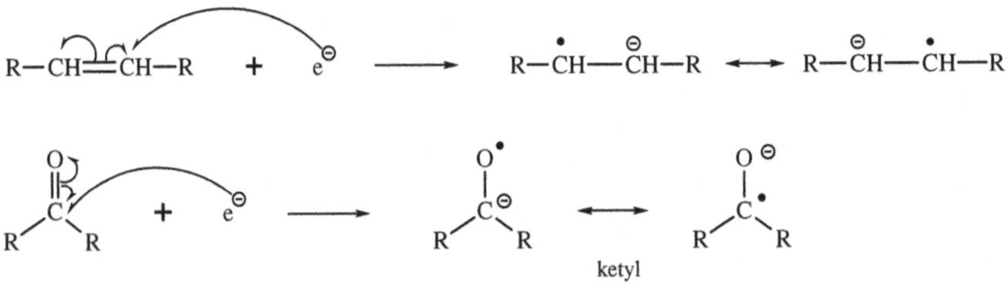

Fig. 10.2

These radical anions are usually reactive intermediates, but can be stabilized by delocalization. For example, the addition of an electron to benzophenone produces a purple radical anion, known as benzophenone ketyl, which is long lived (or persistent) in the absence of protons because the unpaired electron can be delocalized around the two benzene rings (Fig. 10.3). As oxygen is more electronegative than carbon, the negative charge density is situated mainly on the oxygen atom and the unpaired electron is located primarily on the carbon atom.

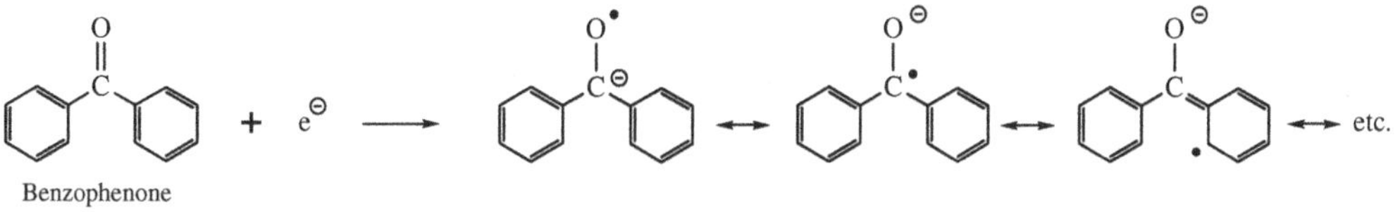

Benzophenone

Fig. 10.3

The most common method of forming radical anions involves the reduction of aromatics or carbonyls with metals, including sodium or potassium, which are able to donate an electron (Fig. 10.4). Conjugated molecules with low-energy LUMOs will readily accept an electron because the radical anions are stabilized by resonance. Therefore an α,β-unsaturated carbonyl is more easily reduced than an isolated carbonyl, and naphthalene is easier to reduce than benzene.

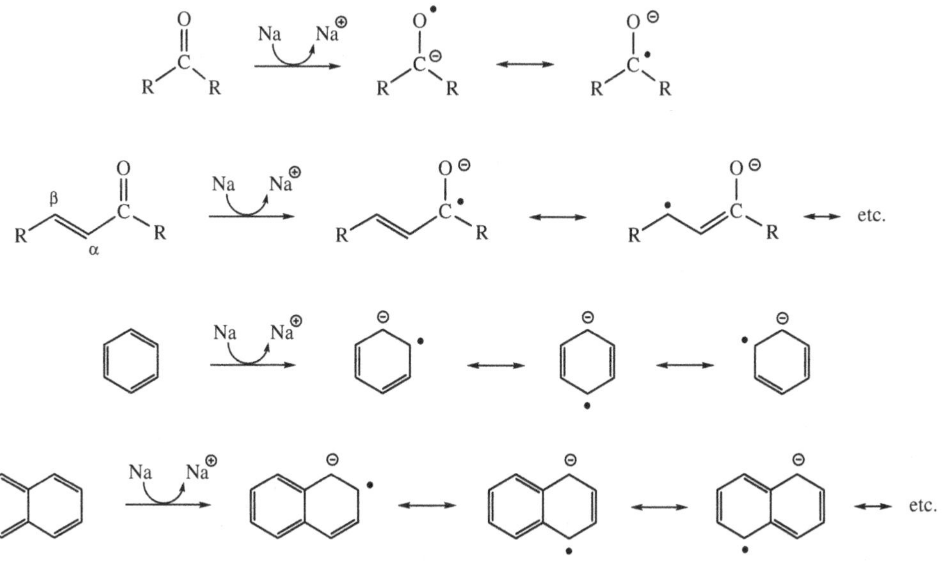

Fig. 10.4

Alternatively, radical anions can be prepared by electrochemical reduction at the cathode or by the addition of a radical to an anion. Thus, the enolate anion derived from propanone is known to add to the phenyl radical to give a radical anion (Fig. 10.5).

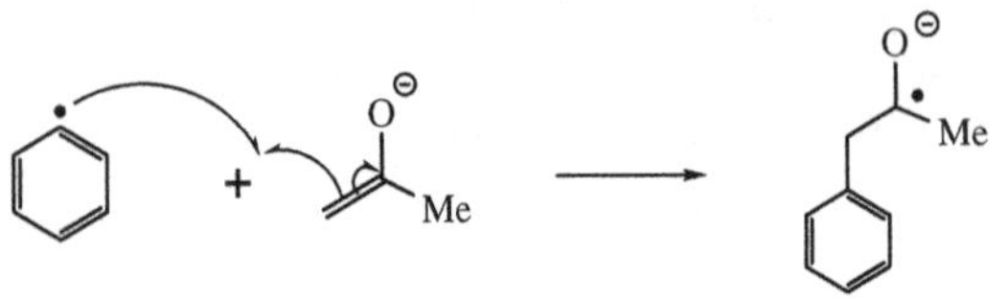

Fig. 10.5

Once formed, radical anions can undergo a variety of different reactions as highlighted in Fig. 10.6. The unpaired electrons of two radical anions can combine to form a new two-electron bond in a dimerization reaction. Alternatively, a disproportionation reaction can take place in which one radical anion loses and another gains an electron. Radical anions can also fragment so as to form the most stable radical and anion products, whilst oxidation can proceed to give a neutral non-radical product. Finally, the negative charge of the radical anion can be neutralized by reaction with a proton to generate a radical which can then react further.

$RA^{\ominus\bullet} + RA^{\ominus\bullet} \longrightarrow [RA{-}RA]^{2\ominus}$	Dimerization
$RA^{\ominus\bullet} + RA^{\ominus\bullet} \longrightarrow RA^{2\ominus} + RA$	Disproportionation (electron transfer)
$RA^{\ominus\bullet} \longrightarrow R^{\bullet} + A^{\ominus}$	Fragmentation (or mesolytic cleavage)
$RA^{\ominus\bullet} + Ox \longrightarrow RA + Ox^{\ominus\bullet}$	Oxidation (loss of an electron)
$RA^{\ominus\bullet} + H^{\oplus} \longrightarrow RAH^{\bullet}$	Protonation

RA = radical anion; Ox = oxidizing agent

Fig. 10.6

Radical anions therefore undergo some characteristic radical reactions (such as dimerization) and some typical anionic reactions (such as protonation). The particular fate of the radical anion will depend on the reaction solvent, the concentration and the presence of oxidizing agents or acids.

10.2 Reactions of radical anions

10.2.1 The pinacol and McMurry reactions

The reduction of ketones to give radical anions forms the basis of the pinacol and McMurry reactions. When acetone is reacted with sodium in aprotic solvents, such as tetrahydrofuran or benzene, the intermediate (ketyl) radical anion can couple to form a dianion (167) (Fig. 10.7). In the process, a strong carbon–carbon bond is formed and this is the driving force for the reaction. The rate of the coupling is expected to be slower than that for radical reactions, because the radical anions are negatively charged and therefore subject to electrostatic repulsion. Dianion (167) can be protonated to give 2,3-dimethylbutane-2,3-diol (168), which is also known as pinacol, hence the name of the reaction. Couplings can be carried out using other ketones and, even though these do not produce pinacol (but related diols), the same name is used.

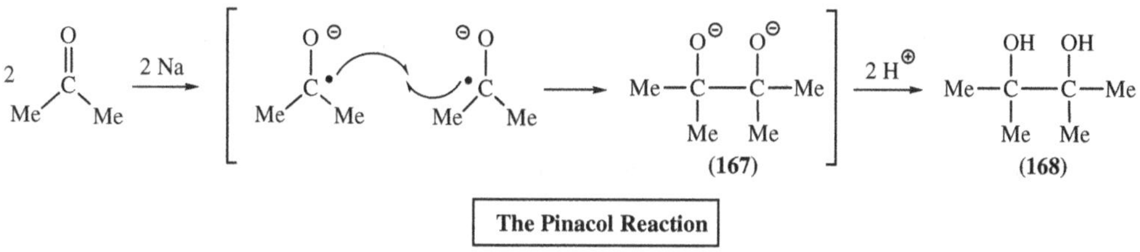

The Pinacol Reaction

Fig. 10.7

The reaction often gives better yields when magnesium is used as the reducing agent, because the magnesium ion (generated on electron transfer) can form two strong covalent bonds to two ketyl oxygens (Fig. 10.8). As a consequence, the ketyls (which are held in close proximity) behave more like radicals and couple more efficiently to give the five-membered ring (169), which can be hydrolysed to give pinacol in approximately 45% yield.

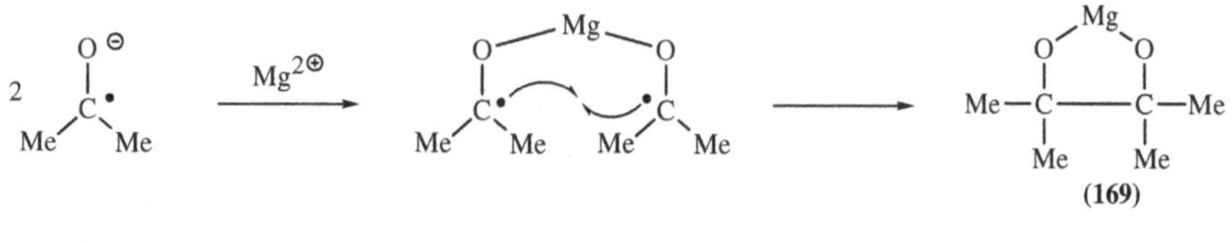

Fig. 10.8

191

Other reducing agents can also be used, and samarium(II) iodide, for example, has been shown to promote the coupling of benzaldehyde to produce the diol **(170)** in excellent yield (Fig. 10.9).

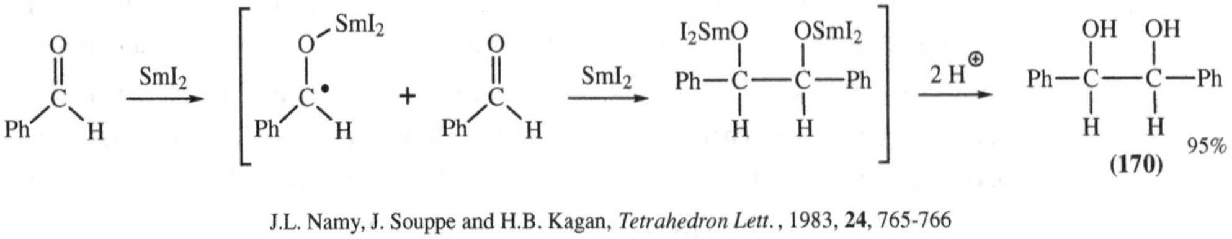
SmI2
SmI2
SmI2
I2SmO
OSmI2
2 H⊕
95%
(170)
J.L. Namy, J. Souppe and H.B. Kagan, Tetrahedron Lett., 1983, 24, 765-766

Fig. 10.9

If titanium metal is used, an alkene rather than a diol is usually produced in a McMurry reaction (Fig. 10.10). The titanium metal, generated by reduction of titanium(III) (using K, Li or $LiAlH_4$), transfers an electron to the carbonyl, and the resulting ketyl couples to give the dianion **(171)**. At low temperatures, this dianion can be protonated to give the diol, but usually deoxygenation occurs, through binding to the surface of the titanium metal, to form an alkene. The mechanism of deoxygenation is not completely understood, although initial homolytic cleavage of one of the O—C bonds could lead to the alkene and titanium dioxide (TiO_2) products.

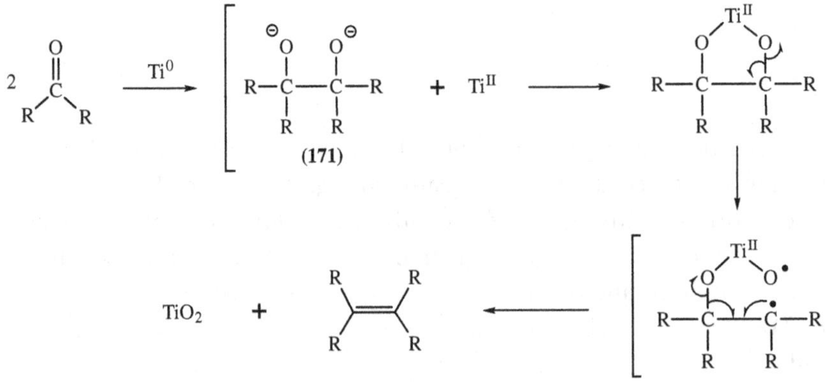
Ti0
(171)
+ TiII
TiII
TiII
TiO2

Fig. 10.10

For these coupling reactions to take place, an aprotic solvent must be used. If a ketone is reduced by sodium in a protic solvent, such as ethanol, the ketyl (radical anion) can undergo a competitive protonation reaction (Fig. 10.11). Protonation can occur before coupling takes place, which produces a radical that can be reduced by a further equivalent of sodium. On protonation, an alcohol rather than a diol is formed.

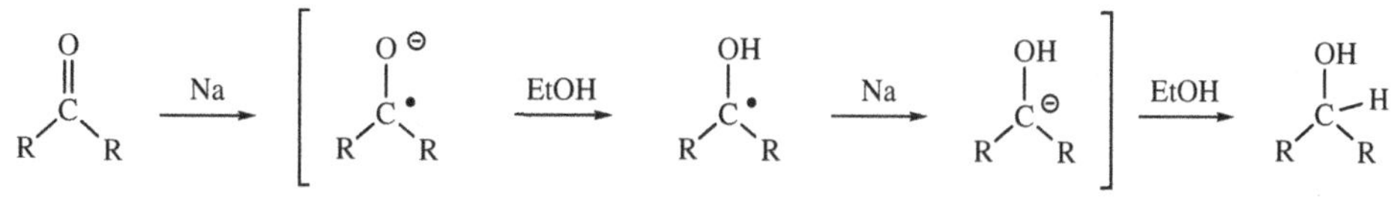

Fig. 10.11

10.2.2 The acyloin condensation

Esters can also be reduced to form radical anions, which can couple to produce hydroxyketones or acyloins (Fig. 10.12). Reaction of an ester with sodium in solvents such as xylene is presumed to form an initial dianion **(172)** which possesses two alkoxy leaving groups (cf. the pinacol reaction). Elimination of these leaving groups produces a diketone that can be reduced further, and addition of two electrons produces the enediolate **(173)**. Finally, on protonation, a 1,2-enediol is formed and this tautomerizes to give the more stable hydroxyketone **(174)**.

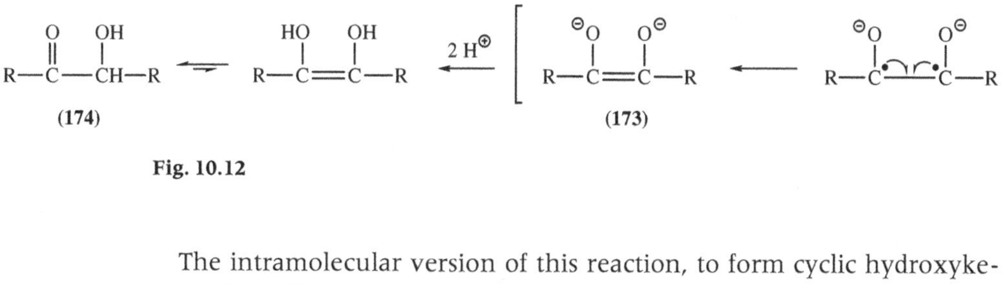

Fig. 10.12

The intramolecular version of this reaction, to form cyclic hydroxyketones from diesters, is particularly effective; the reaction does not require high dilution, which is often employed in other methods of cyclization (to minimize competing intermolecular reactions) (Fig. 10.13). This has been explained by the sodium surface acting as a template, which holds the radical centres close together leading to an efficient cyclization. Even large rings (>10 membered) can be prepared in good yield (typically 60–95%), and yields are often improved by adding trimethylsilyl chloride (Me_3SiCl)

to the reaction. The Me₃SiCl traps the reactive enediolate (175) as soon as it is formed and prevents this dianion from undergoing side reactions. The resulting bis-silyl enol ether (176) can be hydrolysed to give the acyloin.

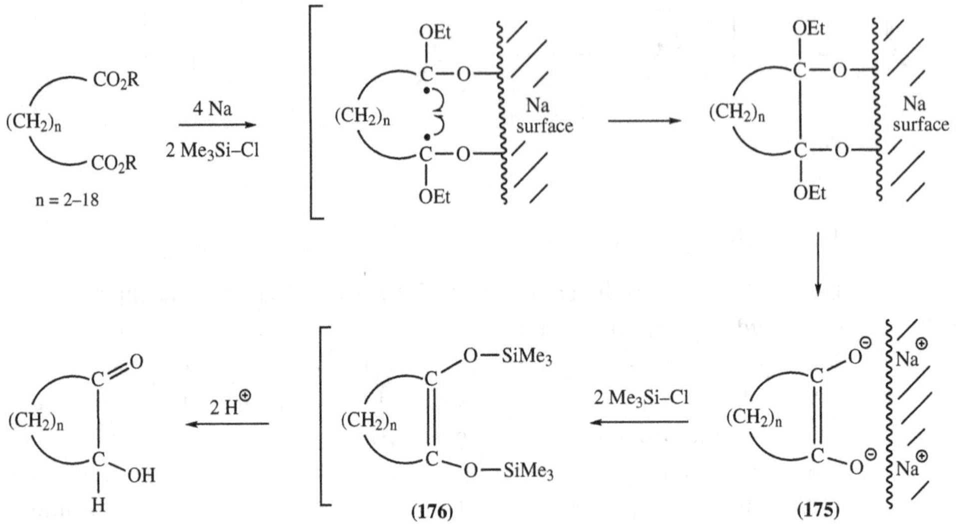

The Acyloin Condensation

Fig. 10.13

Like the pinacol reaction, the acyloin condensation reaction requires an aprotic solvent. If protic solvents are used, the ketyl (radical anion) will be protonated and the ester will be reduced to give alcohols (in the Bouveault–Blanc reaction) (Fig. 10.14).

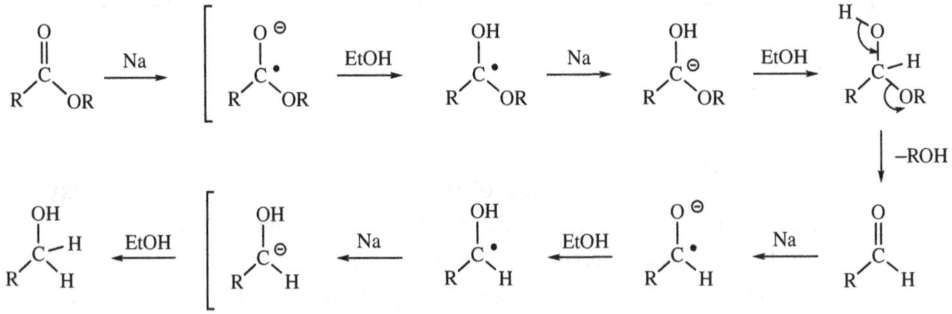

The Bouveault–Blanc Reaction

Fig. 10.14

10.2.3 The Birch reduction

The addition of an electron to a conjugated alkene produces a radical anion stabilized by delocalization. This is usually carried out under Birch reduction conditions, which involve the use of lithium, sodium or potassium in liquid ammonia, often in the presence of an alcohol. These solutions generate solvated metal cations and electrons, and the small reactive electrons can add to even hindered double bonds. The most common application of this reduction uses aromatic precursors and, for example, benzene is reduced to 1,4-cyclohexadiene (177) (Fig. 10.15). The reaction is believed to involve the addition of an electron to the aromatic ring, producing radical anion (178), which is then protonated by an alcohol. Further reduction leads to anion (179), and this forms diene (177) after a second protonation.

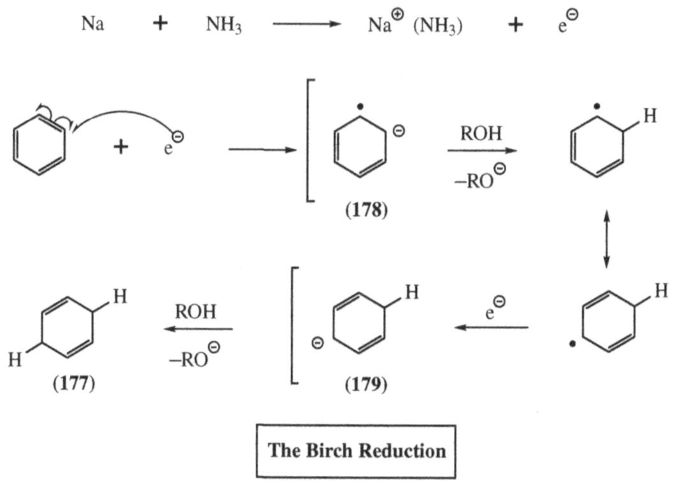

The Birch Reduction

Fig. 10.15

Substituted benzene rings can also be reduced, and the electronic nature of the substituent determines which particular diene (regioisomer) is formed. If an electron-donating substituent (such as OMe) is present, a 1,4-cyclohexadiene is produced, while an electron-withdrawing substituent (such as CO_2^-) gives the 2,5-cyclohexadiene selectively (Fig. 10.16). The different regioselectivity has been explained by the stability of the intermediate radical anions; the negative charge is destabilized when adjacent to an electron-donating group, but stabilized by an adjacent electron-withdrawing group. The benzene ring rather than the carbonyl group accepts the electron because the carboxylate group (of the benzoate salt) is negatively charged. Otherwise, a radical anion with a negative charge on an adjacent oxygen atom would be produced and this would be very unstable.

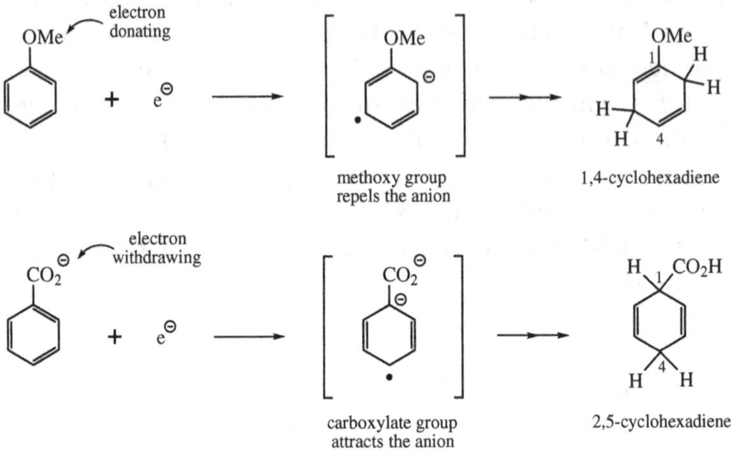

Fig. 10.16

This method of reduction is not only useful for preparing dienes, but is widely used in synthesis for deprotection of benzyl protecting groups. These groups are used to protect amine and, particularly, alcohol functional groups (Fig. 10.17). Benzyl ethers are prepared from alcohols by alkylation of intermediate alkoxides using benzyl bromide or chloride (in the Williamson ether synthesis). Converting the alcohol to a benzyl ether reduces the reactivity, and this functional group is stable to a variety of acidic and basic conditions.

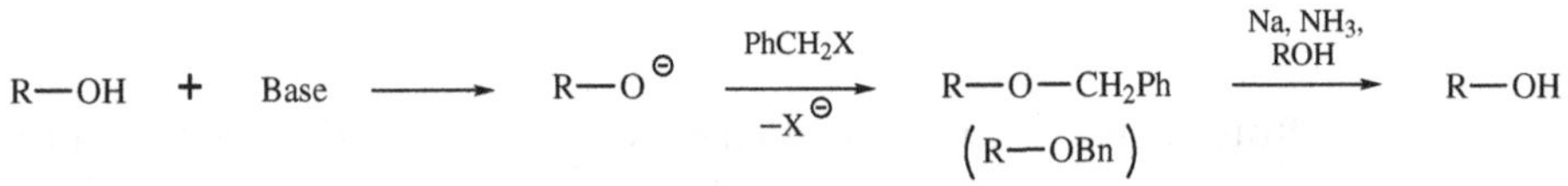

Fig. 10.17

When the alcohol needs to be regenerated, the deprotection can be accomplished using the Birch reduction reaction (Fig. 10.18). Reduction of the aromatic ring produces anion (180), which is a 1,4-diene because the electron-donating alkyl substituent repels the negative charge. This anion is able to fragment with elimination of the alkoxide anion, which is protonated on acidic work-up to afford the (deprotected) alcohol.

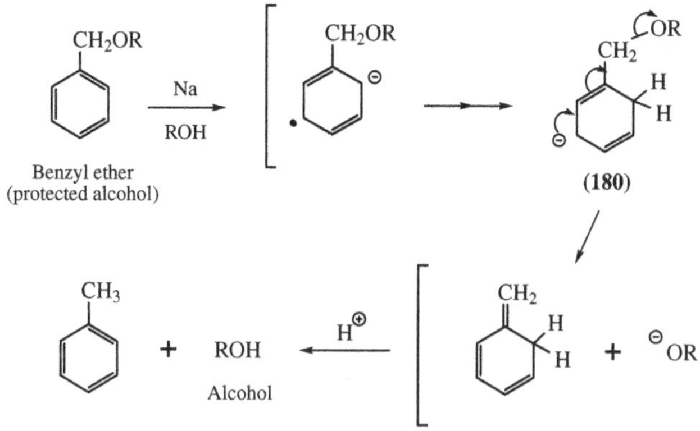

Fig. 10.18

Alkynes can also be converted to *trans*-alkenes using this method of reduction. Whereas hydrogenation of alkynes, using a Lindlar catalyst (comprising Pd–CaCO$_3$–PbO), produces *cis*-alkenes, reaction with sodium in liquid ammonia forms *trans*-alkenes (Fig. 10.19).

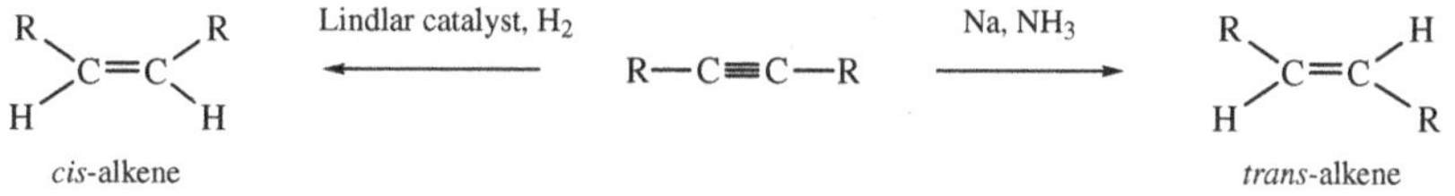

Fig. 10.19

One possible mechanism to explain the *trans* stereoselectivity is shown in Fig. 10.20. Addition of an electron to the alkyne triple bond produces radical anion **(181)**; this is expected to have a *trans* geometry because of electron repulsion between the radical and carbanion orbitals. Protonation of the reactive vinyl carbanion (with ammonia) produces the *trans*-vinyl radical **(182)**. Although vinyl radicals usually equilibrate, the reductions are carried out at low reaction temperature (−33°C) to ensure that *trans* to *cis* isomerization is prohibitively slow. A second reduction step forms the *trans*-vinyl anion **(183)**, with the bulky R groups on opposite faces. Vinyl anions **(183)** equilibrate more slowly than vinyl radicals, and therefore protonation typically affords >98% of the *trans*-alkene **(184)**.

197

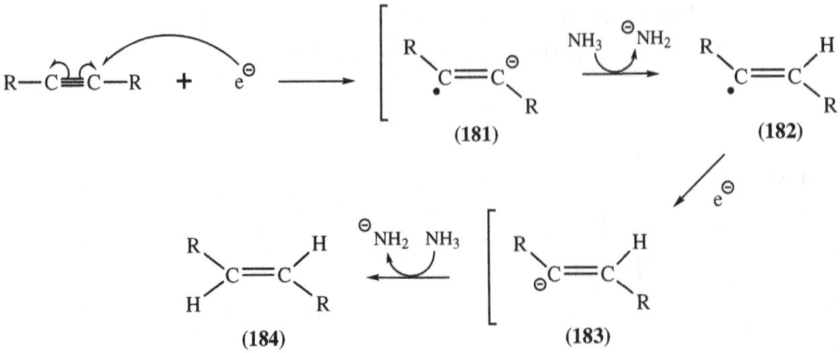

(181) (182)

(184) (183)

Fig. 10.20

10.2.4 Unimolecular radical nucleophilic substitution ($S_{RN}1$)

Nucleophilic substitution of a halide group on a benzene ring is very rare. Nucleophilic aromatic substitution (S_NAr) reactions are usually only possible when the benzene ring contains electron-withdrawing substituents that are able to stabilize the intermediate carbanion which is formed on nucleophilic attack. An alternative and more general approach to S_NAr reactions involves the formation of radical anion intermediates in an $S_{RN}1$ mechanism (Fig. 10.21).

The S_{RN} 1 Mechanism

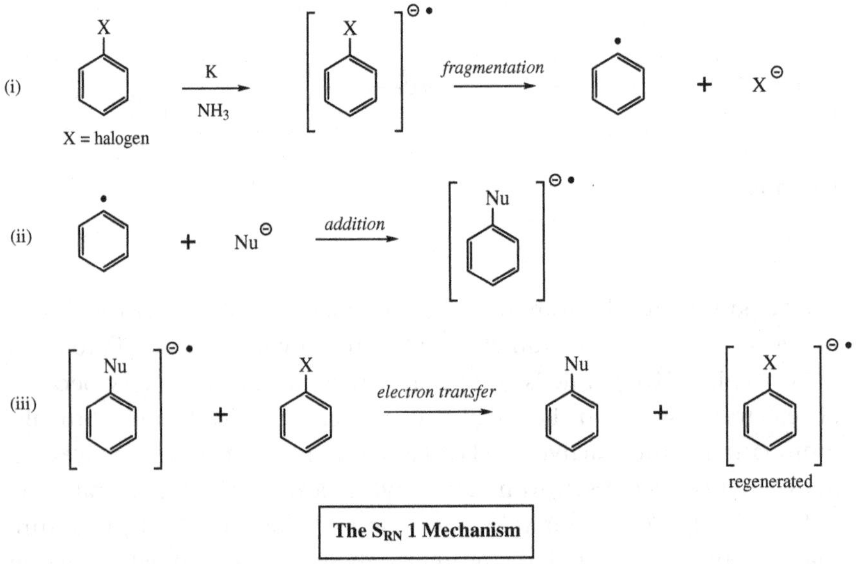

Fig. 10.21

The addition of an electron, generated, for example, from potassium in liquid ammonia, to an aryl halide produces a radical anion that can fragment [step (i)], and this unimolecular step is signified by the number 1 in $S_{RN}1$. An aryl radical (rather than anion) is generated because the negative charge prefers to be associated with the electronegative halogen atom (X). The aryl radical can form a new radical anion by addition to a nucleophile (Nu⁻), typically a carbanion [step (ii)]. The intermediate radical anion can then be converted to the substituted benzene by transfer of an electron to a molecule of aryl halide (starting material) to complete a chain mechanism [step (iii)].

A variety of different (carbon, sulfur or phosphorus) nucleophiles can be used, and the substitutions are regioselective, i.e. the nucleophile is introduced at the same position as the halide group in the starting material (Fig. 10.22).

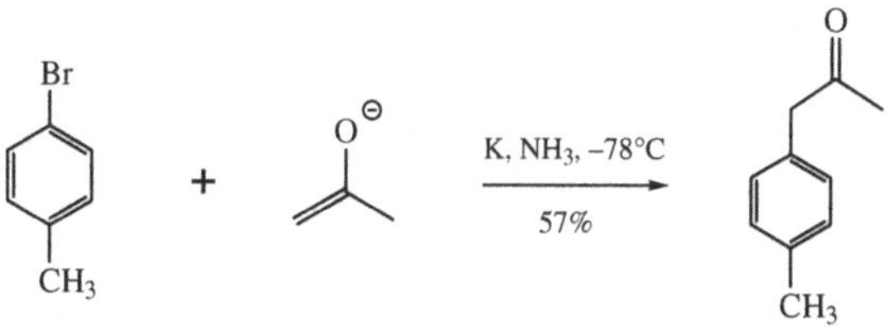

R.A. Rossi and J.F. Bunnett, *J. Am. Chem. Soc.*, 1972, **94**, 683–684

Fig. 10.22

10.2.5 Superoxide—a biological radical anion

Addition of an electron to molecular oxygen produces the superoxide radical anion (Fig. 10.23). This is an important reaction in biological systems where, for example, metalloenzymes can act as electron donors. Although the superoxide anion is not a particularly reactive species, it can be converted to considerably more reactive radicals which can then damage any proteins, fats or deoxyribonucleic acid (DNA) within the cell. On protonation, the peroxyl radical is generated which can abstract a hydrogen atom from biological molecules (RH) to form hydrogen peroxide. The hydrogen peroxide can react further in the presence of an electron donor, and reduction yields the extremely reactive hydroxyl radical. The hydroxyl radical causes most of the structural damage in biological systems, and it can modify DNA (leading to a mutation) by, for example, adding to the constituent bases.

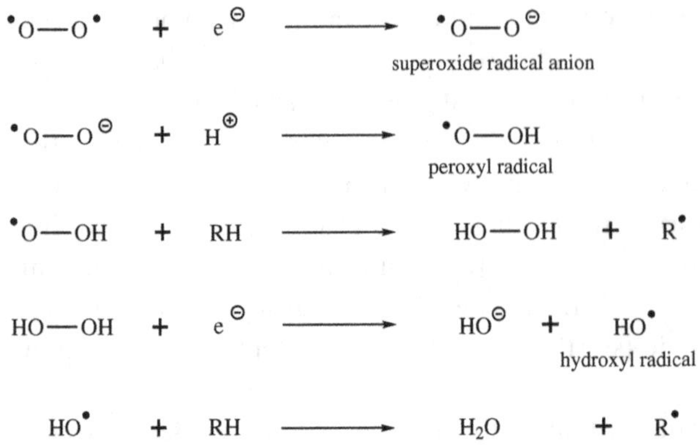

$$\overset{\bullet}{O}-O\overset{\bullet}{} \quad + \quad e^{\ominus} \quad \longrightarrow \quad \overset{\bullet}{O}-O^{\ominus}$$

superoxide radical anion

$$\overset{\bullet}{O}-O^{\ominus} \quad + \quad H^{\oplus} \quad \longrightarrow \quad \overset{\bullet}{O}-OH$$

peroxyl radical

$$\overset{\bullet}{O}-OH \quad + \quad RH \quad \longrightarrow \quad HO-OH \quad + \quad R^{\bullet}$$

$$HO-OH \quad + \quad e^{\ominus} \quad \longrightarrow \quad HO^{\ominus} \quad + \quad HO^{\bullet}$$

hydroxyl radical

$$HO^{\bullet} \quad + \quad RH \quad \longrightarrow \quad H_2O \quad + \quad R^{\bullet}$$

Fig. 10.23

Living cells therefore need to control the level of these damaging radicals, and superoxide dismutase (SOD) enzymes are known to regulate the levels of the superoxide radical anion (Fig. 10.24). These enzymes catalyse a disproportionation reaction which, in the presence of acid, leads to the formation of oxygen and hydrogen peroxide. One of the radical anions loses an electron to regenerate oxygen, whereas the other accepts an electron to form a dianion that is subsequently protonated. The production of reactive hydrogen peroxide may not appear to be very sensible, but catalase or peroxidase enzymes are present in the living cell to destroy the hydrogen peroxide. It is clearly very important that the hydrogen peroxide is rapidly 'neutralized', and hence catalase is one of the most active enzymes in living cells: 6 million molecules of hydrogen peroxide can be transformed by one molecule of the enzyme every minute!

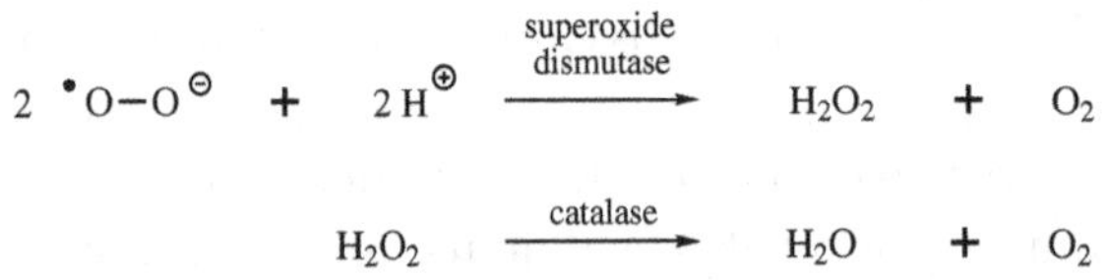

$$2\ \overset{\bullet}{O}-O^{\ominus} \quad + \quad 2\,H^{\oplus} \quad \xrightarrow{\text{superoxide dismutase}} \quad H_2O_2 \quad + \quad O_2$$

$$H_2O_2 \quad \xrightarrow{\text{catalase}} \quad H_2O \quad + \quad O_2$$

Fig. 10.24

10.3 Summary

Radical anions are produced from the addition of an electron to unsaturated molecules, including carbonyls, alkenes and aromatics. These intermediates, which are common in reductions using metals (or electrolysis), can undergo reactions typical of radicals or anions. For example, a ketyl (radical anion) can undergo a radical coupling reaction or, alternatively, an anionic protonation reaction.

Further reading

Rossi, R.A. (1982) Phenomenon of radical anion fragmentation in the course of aromatic $S_{RN}1$ reactions. *Accounts of Chemical Research*, **15**, 164–170.

Savéant, J.-M. (1990) Single electron transfer and nucleophilic substitution. *New Journal of Chemistry*, 1992, **16**, 131–150.

Sawyer, D.T. & Valentine, J.S. (1981) How super is superoxide? *Accounts of Chemical Research*, **14**, 393–400.

CHAPTER 11

Radical Cations

11.1 Introduction

When a neutral (non-radical) molecule loses an electron, a species with both a positive charge and an unpaired electron is formed. This radical cation can have the unpaired electron and positive charge located on the same atom, or on different atoms (these are called distonic species). The electron is removed from the highest occupied molecular orbital (HOMO) of the precursor, and the higher the energy of this orbital, the more easily the electron is released (Fig. 11.1). Compounds with a lone pair or π bond(s) can readily lose an electron because the non-bonding or π bond orbitals are relatively high in energy (Fig. 11.2). In comparison, removal of an electron from a lower energy σ bond is much more difficult, and the ease with which a molecule will lose an electron can be determined from its ionization potential; the lower the ionization potential, the more easily a radical cation is formed.

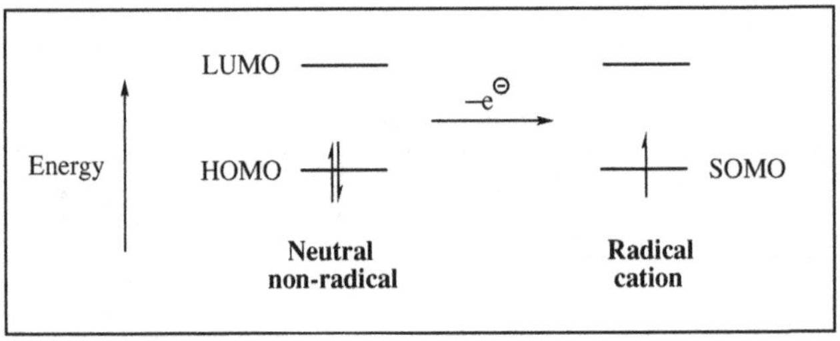

Fig. 11.1

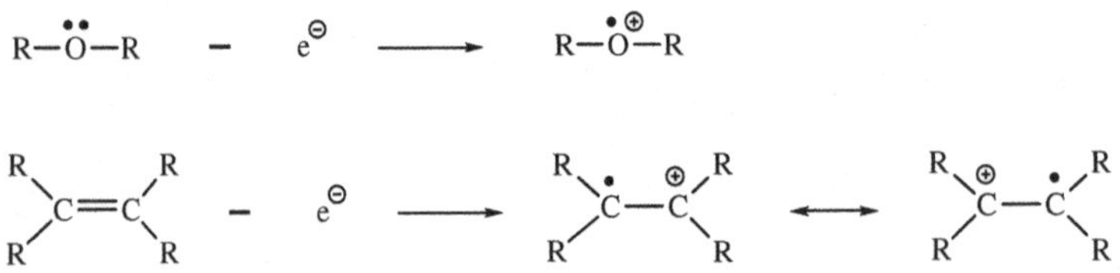

Fig. 11.2

Radical cations are generally very reactive as the loss of an electron, particularly a bonding electron, weakens the molecular structure. However, they can be stabilized by delocalization of the unpaired electron, and Wurster's salts, for example, can even be isolated (Fig. 11.3).

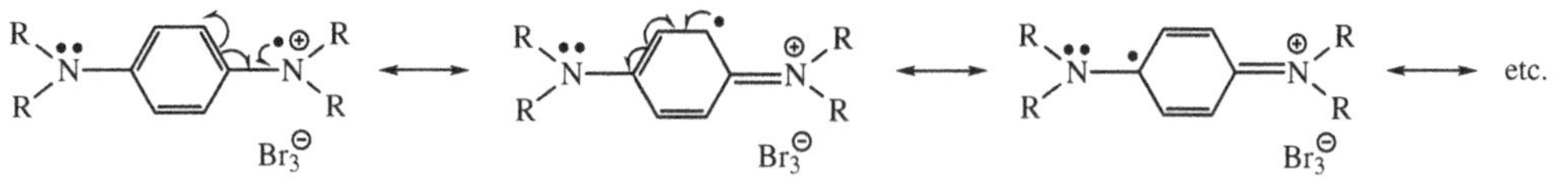

Wurster's salts (R=H or Me)

Fig. 11.3

There are a number of different methods of oxidation which can be used to prepare radical cations. Oxidizing metal ions [such as manganese(III), cobalt(III), lead(IV), cerium(IV) or silver(II)] will readily accept an electron, and aromatic radical cations can be generated using cobalt(III) salts (Fig. 11.4). The presence of electron-donating substituents on the benzene ring stabilizes the radical cation, which has five π electrons, and thereby increases the rate of oxidation.

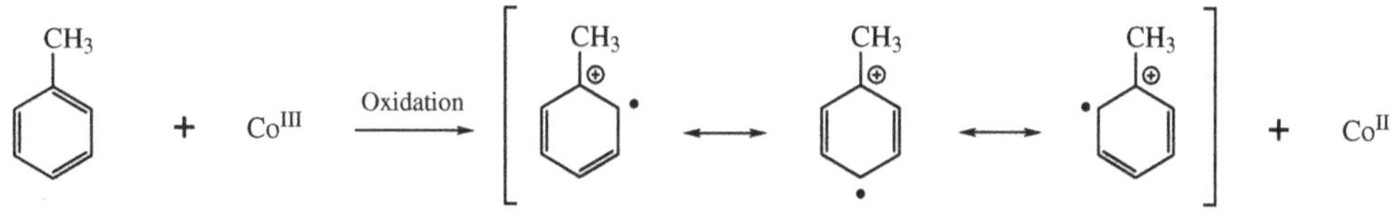

Fig. 11.4

Electrochemical oxidation at the anode is also possible and this is commonly used to oxidize phenols (Fig. 11.5).

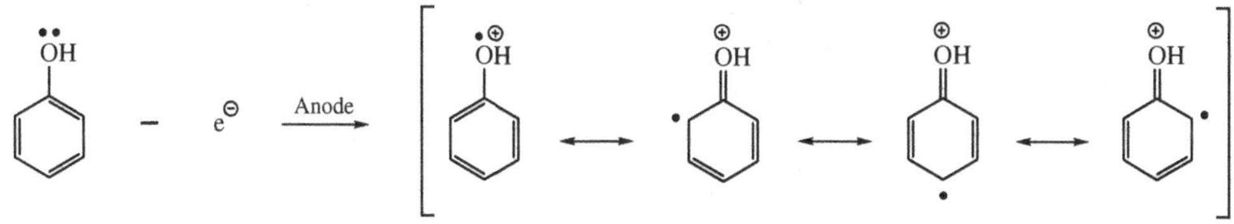

Fig. 11.5

Radical cations are also generated by electron impact in mass spectrometry (Fig. 11.6). Molecules are bombarded with high-energy electrons and ionization forms a radical cation known as the molecular (or parent) ion. In a mass spectrum, the molecular ion peak is usually the peak of highest mass number and corresponds to the molecular weight of the compound.

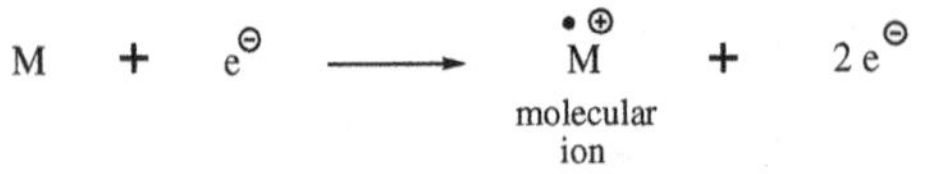

$$M \quad + \quad e^{\ominus} \quad \longrightarrow \quad \overset{\bullet\,\oplus}{M} \quad + \quad 2\,e^{\ominus}$$
molecular
ion

Fig. 11.6

Alternatively, radical cations can be formed by addition of a radical to a cation and, for example, nucleophilic alkyl radicals will generate radical cations on reaction with protonated (and electrophilic) pyridine rings (Fig. 11.7).

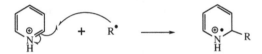

Fig. 11.7

Once formed, radical cations can undergo the various reactions highlighted in Fig. 11.8. The unpaired electrons of two radical cations can combine to form a dication, or a disproportionation reaction can lead to one radical cation gaining an electron and the other losing an electron. Fragmentations can proceed to form the most stable radical and cation products, and this process is commonly observed in mass spectrometry. The molecular ion generally decomposes and the mass of the positively charged fragment provides useful structural information. Reduction of a radical cation by addition of an electron produces a neutral non-radical product (and a new radical cation), while both alkenes and nucleophiles will add to radical cations. Hydrogen atom abstraction reactions can also take place, particularly when the abstraction can occur intramolecularly. Radical cations therefore undergo some characteristic radical reactions (such as coupling and hydrogen atom abstraction) and some typical cationic reactions (such as nucleophilic addition). They are generally more reactive than the corresponding radical anions and are intermediates in a number of organic reactions.

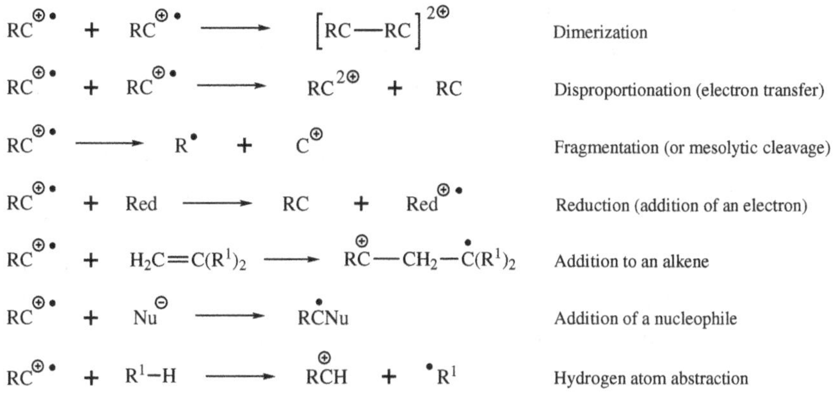

$$RC^{\oplus\bullet} + RC^{\oplus\bullet} \longrightarrow \left[RC{-}RC\right]^{2\oplus} \qquad \text{Dimerization}$$

$$RC^{\oplus\bullet} + RC^{\oplus\bullet} \longrightarrow RC^{2\oplus} + RC \qquad \text{Disproportionation (electron transfer)}$$

$$RC^{\oplus\bullet} \longrightarrow R^\bullet + C^\oplus \qquad \text{Fragmentation (or mesolytic cleavage)}$$

$$RC^{\oplus\bullet} + Red \longrightarrow RC + Red^{\oplus\bullet} \qquad \text{Reduction (addition of an electron)}$$

$$RC^{\oplus\bullet} + H_2C{=}C(R^1)_2 \longrightarrow RC{-}CH_2{-}\overset{\bullet}{C}(R^1)_2 \qquad \text{Addition to an alkene}$$

$$RC^{\oplus\bullet} + Nu^\ominus \longrightarrow \overset{\bullet}{RC}Nu \qquad \text{Addition of a nucleophile}$$

$$RC^{\oplus\bullet} + R^1{-}H \longrightarrow \overset{\oplus}{RC}H + {}^\bullet R^1 \qquad \text{Hydrogen atom abstraction}$$

RC = radical cation; Red = reducing agent; Nu = nucleophile; R^1 = alkyl group

Fig. 11.8

11.2 Reactions of radical cations

11.2.1 Benzylic oxidation

Electron-rich aromatic rings are readily oxidized to radical cations using metals such as manganese(III) or cerium(IV). When a methyl side chain is present on the benzene ring, the radical cation can fragment (by loss of a proton) to form a resonance-stabilized benzylic radical (Fig. 11.9). The radical can be oxidized by a second equivalent of the metal ion to give a delocalized benzylic cation, which can react with a nucleophile (Nu). The nucleophile attacks the benzylic carbon to form a stable product containing an aromatic benzene ring.

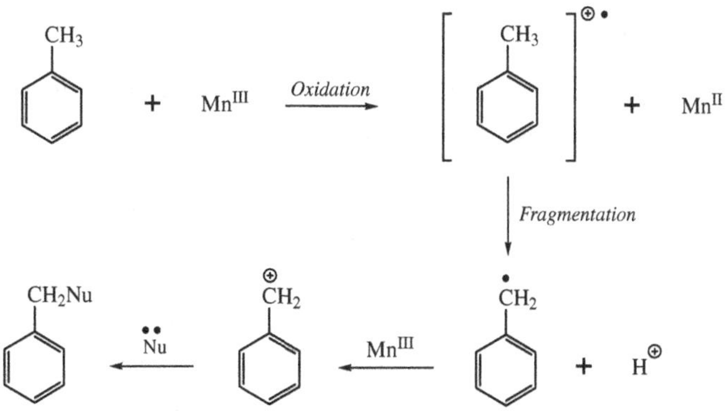

Fig. 11.9

These reactions can therefore be used to selectively oxidize (and functionalize) the benzylic position, which has found application in the deprotection of *para*-methoxybenzyl (PMB) ethers (Fig. 11.10). Alcohols are
often protected as PMB ethers using a Williamson reaction (see Section
10.2.3), and selective oxidative deprotection can be carried out using
ammonium cerium(IV) nitrate $[(NH_4)_2Ce^{IV}(NO_3)_6]$. Electron transfer to
cerium(IV) produces an initial radical cation **(185)**, and fragmentation
leads to a benzylic radical **(186)**, which is oxidized by a second equivalent
of cerium(IV). The resultant benzylic cation **(187)** is trapped by reaction
with water to give a hemiacetal **(188)**, which decomposes to *para*-
methoxybenzaldehyde and the desired 'deprotected' alcohol.

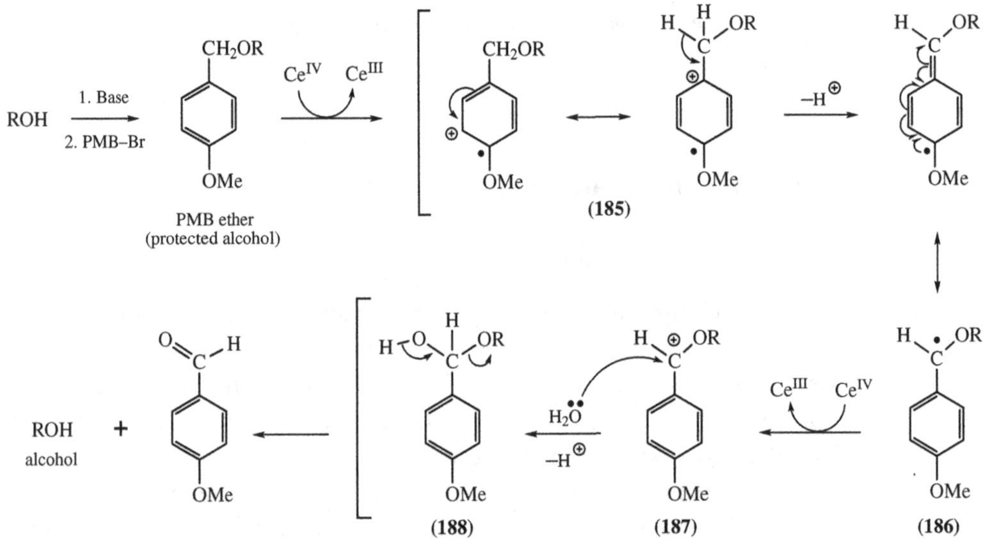

Fig. 11.10

11.2.2 Oxidative phenolic coupling

Phenols are readily oxidized by a variety of one-electron oxidants to form
radical cations (Fig. 11.11). These rapidly fragment (by loss of a proton) to
form phenoxyl (or aryloxyl) radicals that are stabilized by resonance. This
reaction is very important in nature, as a number of alkaloid (see Section
7.6) and lignan natural products are derived from coupling of these radicals. These reactions are initiated by peroxidase enzymes (which accept
an electron), and the phenoxyl radicals couple at oxygen or the C-2 (*ortho*)
or C-4 (*para*) positions of the benzene ring to make a new C—O or C—C
bond. Coupling of two oxygen-centred radicals is not observed, as this
would produce a dimer (peroxide) with a very weak O—O bond.

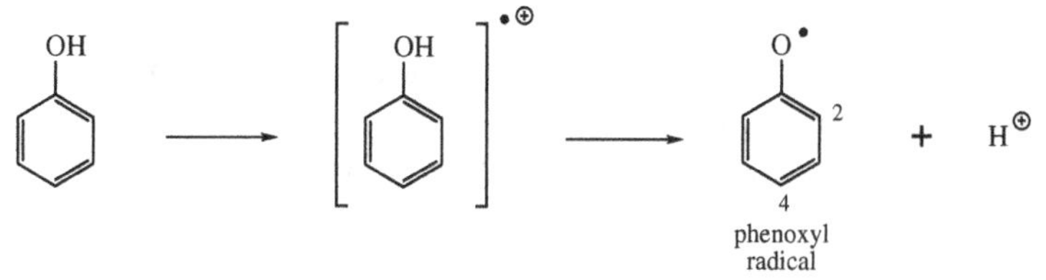

Fig. 11.11

A phenolic coupling has been proposed as the key step in the biosynthesis of the antifungal agent griseofulvin **(189)** (Fig. 11.12). The spirocyclic ring system is thought to arise from the coupling of diradical **(191)**, which is generated on oxidation of diphenol **(190)**. This mechanism has been supported by the formation of racemic dehydrogriseofulvin **(192)** [from **(190)**] using potassium ferricyanide, $K_3Fe(CN)_6$, a well-known, one-electron oxidant.

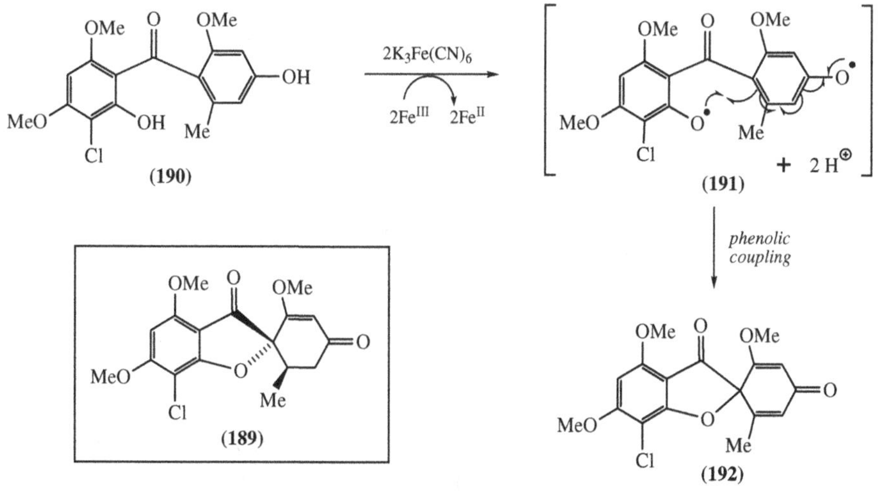

Fig. 11.12

11.2.3 Substitution of heteroaromatics

The formation of radical cations by addition of radicals to cations has been used to alkylate selectively heteroaromatic compounds such as pyridine. At low pH, the pyridine nitrogen is protonated and nucleophilic alkyl radicals will add selectively to the 2- and/or 4-position of the pyridinium ring (Fig. 11.13). The intermolecular addition is facilitated by protonation as nucleophilic radicals will add more rapidly to the electron-poor heterocyclic ring. Deprotonation of the resulting radical cation leads to an azacyclohexadienyl radical, (193) or (194), which in the presence of an oxidizing agent can be oxidized to give the substituted pyridine.

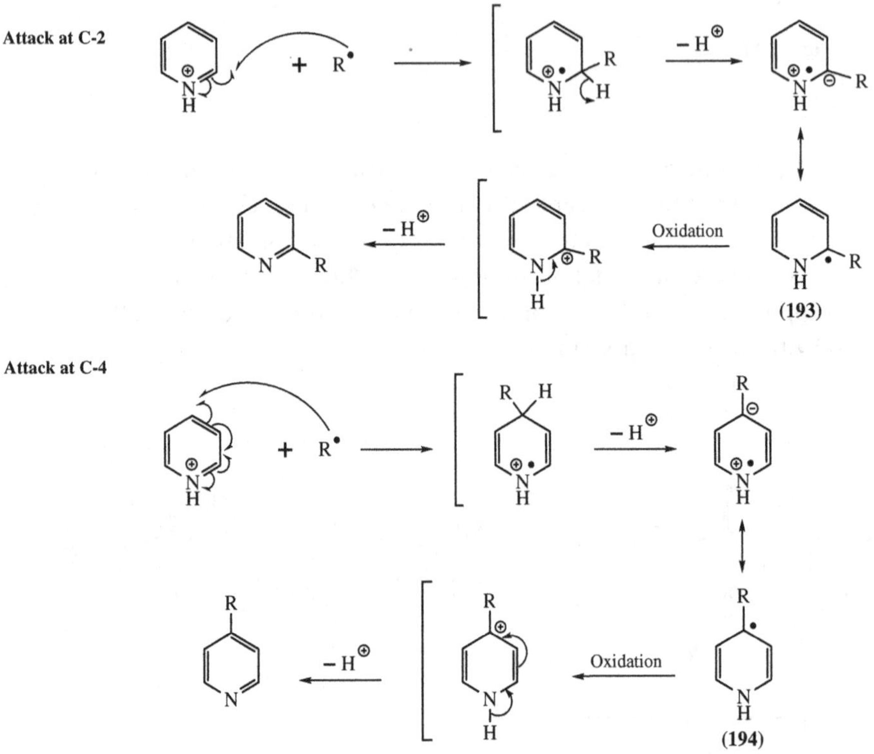

Fig. 11.13

The radicals are usually generated from reaction of carboxylic acids with sodium peroxydisulfate ($Na_2S_2O_8$) and a catalytic amount of silver(I). Peroxydisulfates are very strong oxidizing agents and addition of silver(I) will produce very reactive silver(II). The silver(II) oxidizes the carboxylic acid to generate a radical cation [and silver(I)], which loses a proton, and the resulting carboxyl radical generates an alkyl radical on decarboxylation (Fig. 11.14). This method requires only a catalytic amount of expensive silver(I) as the peroxydisulfate oxidizes and reoxidizes silver(I) to silver(II). The silver(II) also oxidizes the azacyclohexadienyl radical [(193) or (194), Fig. 11.13] to the substituted pyridine.

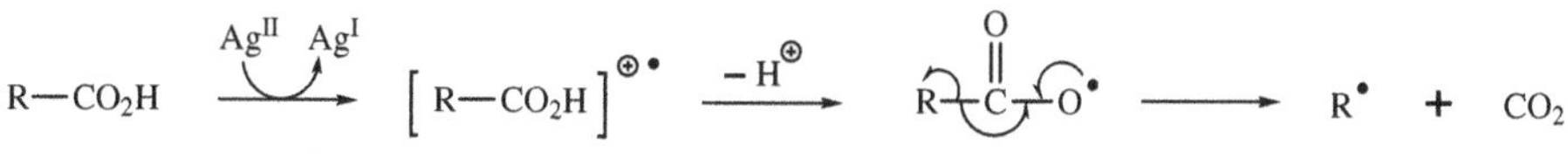

$$R-CO_2H \xrightarrow{Ag^{II} \quad Ag^{I}} \left[R-CO_2H \right]^{\oplus \bullet} \xrightarrow{-H^{\oplus}} R-\overset{\overset{\displaystyle O}{\|}}{C}-O^{\bullet} \longrightarrow R^{\bullet} + CO_2$$

Fig. 11.14

Alkylation of heteroaromatic bases can be achieved using a variety of primary, secondary or tertiary radicals and, for example, isoquinoline **(195)** can be efficiently and regioselectively alkylated by reaction with cyclohexanecarboxylic acid (Fig. 11.15). This method of alkylation is useful because it contrasts with ionic alkylations (i.e. Friedel–Crafts reactions) which work much better with electron-rich rather than with electron-poor aromatic rings.

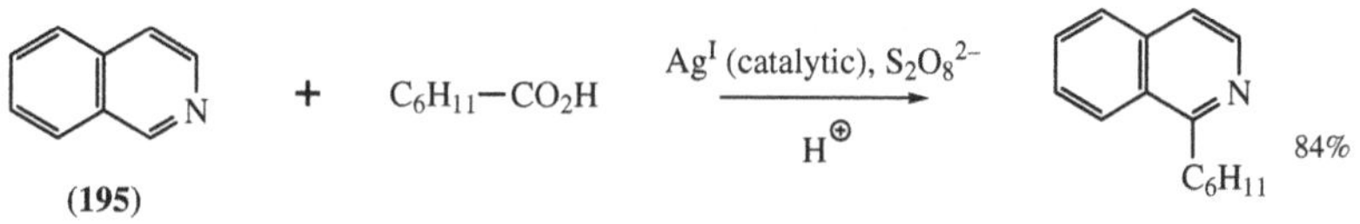

(195) $\quad + \quad C_6H_{11}-CO_2H \quad \xrightarrow[H^{\oplus}]{Ag^I \text{ (catalytic), } S_2O_8^{2-}}$

84%

C_6H_{11}

F. Minisci, R. Bernardi, F. Bertini, R. Galli and M. Perchinummo, *Tetrahedron*, 1971, **27**, 3575–3579

Fig. 11.15

11.2.4 The Hoffmann–Löffler–Freytag reaction

Abstraction of a hydrogen atom by a radical cation is central to the synthesis of pyrrolidine rings using the Hoffmann–Löffler–Freytag reaction (Fig. 11.16). The reaction involves the generation of a nitrogen-centred radical cation **(197)**, called an aminium radical, by homolysis (on heating or irradiation) of the N—Cl bond of a protonated chloroamine **(196)**. The aminium radical can then undergo a 1,5-hydrogen atom transfer to give a distonic radical cation **(198)**, in which the radical and cation groups are well separated. The radical cation **(198)** abstracts a chlorine atom from another molecule of chloroamine **(196)**, in a chain propagating step, to form the protonated chloroamine **(199)** together with radical cation **(197)**. The protonated chloroamine **(199)** is subjected to a basic work-up, and the resulting secondary amine **(200)** cyclizes to form the five-membered pyrrolidine ring **(201)**. The preference for a pyrrolidine ring results from the regioselective hydrogen atom abstraction reaction, which proceeds via a favoured six-membered transition state (with a near-linear N—H—C geometry) (see Chapter 9).

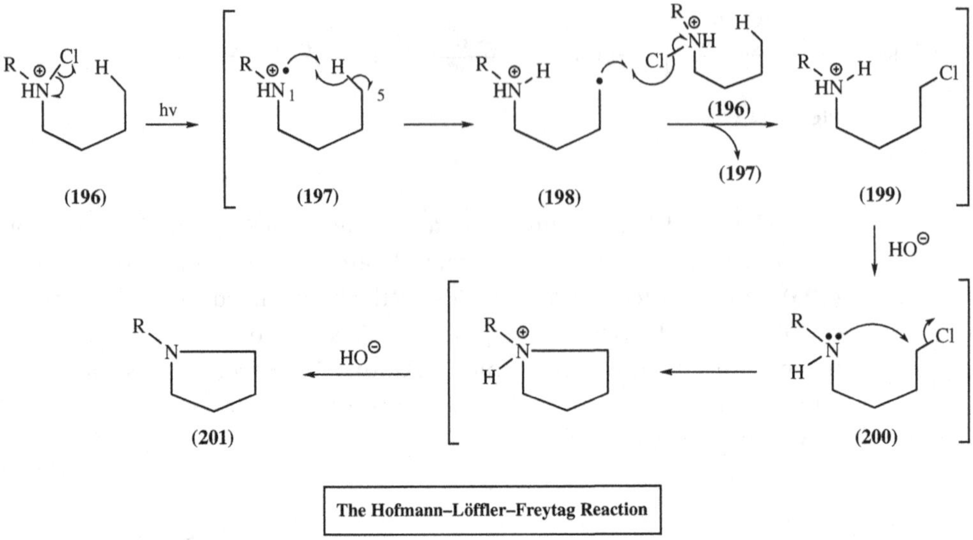

(196)   (197)   (198)   (199)

(201)   (200)

The Hofmann–Löffler–Freytag Reaction

Fig. 11.16

Related hydrogen atom abstraction rearrangements can be observed in mass spectrometry. Carbonyl compounds, for example, can undergo a McLafferty rearrangement that results in the elimination of an alkene fragment (Fig. 11.17). The molecular ion undergoes a 1,5-hydrogen atom abstraction, and expulsion of the alkene gives a charged enol that accounts for a prominent and characteristic peak in the mass spectrum.

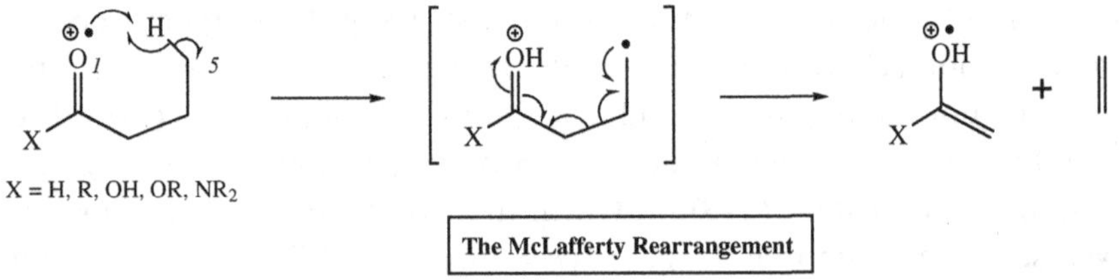

X = H, R, OH, OR, NR$_2$

The McLafferty Rearrangement

Fig. 11.17

11.2.5 The herbicide paraquat

Paraquat **(202)** is a herbicidal bipyridinium salt that is sprayed between the rows of crops to kill weeds selectively. The biological activity of this compound arises from its ability to interfere with the photosynthetic process in the plant by readily accepting an electron. Reduction of paraquat **(202)** is very facile because addition of an electron produces radical cation

(203), which is stabilized by delocalization over the extended conjugated system (Fig. 11.18). However, addition of the electron is reversible and **(203)** can be oxidized back to **(202)** by molecular oxygen. This process produces the superoxide radical anion ($O_2^{-\bullet}$) which helps to cause the rapid death of the plant by forming other toxic oxygen-centred radicals (see Section 10.2.5).

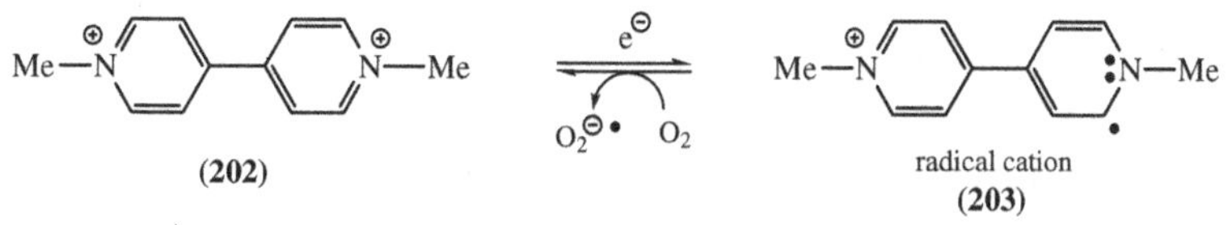

Fig. 11.18

11.3 Summary

Radical cations are generally produced on removal of an electron (i.e. ionization) from a heteroatom lone pair or a π bond. These intermediates, which are formed in oxidation reactions, can fragment to form radicals (generally) by loss of a proton. Radical cations can undergo reactions typical of both radicals and cations. Like carbocations, they will react with nucleophiles and, like radicals, they can undergo hydrogen atom abstraction reactions.

Further reading

Bard, A.J., Ledwith, A. & Shine, H.J. (1976) Formation, properties and reactions of cation radicals in solution. *Advances in Physical Organic Chemistry*, **13**, 155–278.

Hammerich, O. & Parker, V.D. (1984) Kinetics and mechanisms of reactions of organic cation radicals in solution. *Advances in Physical Organic Chemistry*, **20**, 55–189.

Questions

A number of questions have been included to help consolidate your understanding of radical chemistry. These are divided into two sections. The questions in Part 1 are based on the topics discussed in Chapters 1–4 while those in Part 2 cover the material in the remaining Chapters (5–11). Outline answers are also provided which, although not comprehensive, are written so as to cover the key points of each question (based on our current understanding). In order to access more detailed information, a number of references to the primary literature are also included.

Part 1 (based on Chapters 1–4)

1. What radicals will be formed on photolysis or thermolysis of compounds **(204)–(207)**? Show the mechanism of formation of all the possible radicals using curly arrows.

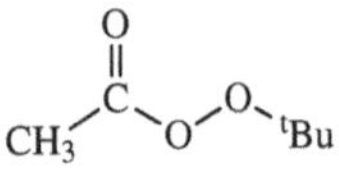

| (204) | (205) | (206) | (207) |

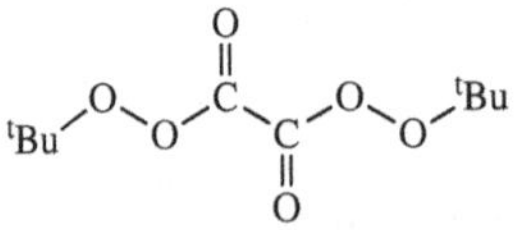

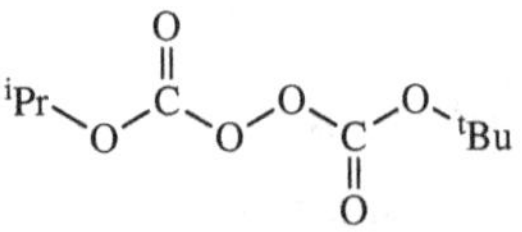

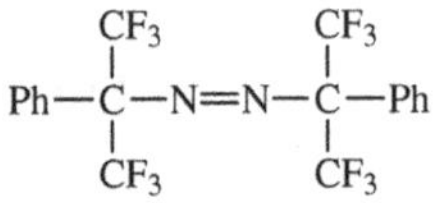

2. What radicals are formed from the following reactions? Do these reactions involve oxidation or reduction of the organic starting material?

(i) $^tBuOO^tBu$ + $FeCl_2$

(ii) $PhCH_2CH(CO_2Et)_2$ + $Mn(OAc)_3$

(iii) $PhN_2^+ Cl^-$ + $CuCl$

(iv) $EtCO_2K$ + Anode

(v) 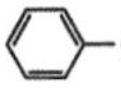   ⟨benzene⟩–OH + $(NH_4)_2Ce(NO_3)_6$

(vi) CH_3O–⟨benzene⟩–CH_3 + $Na_2S_2O_8$ + $AgNO_3$ (catalytic)

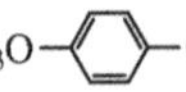

3. Explain why natural products **(208)**–**(210)** react very rapidly with radicals and hence are known as 'free-radical scavengers'.

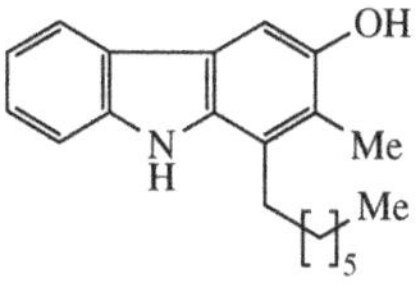

(208)
Carazostatin

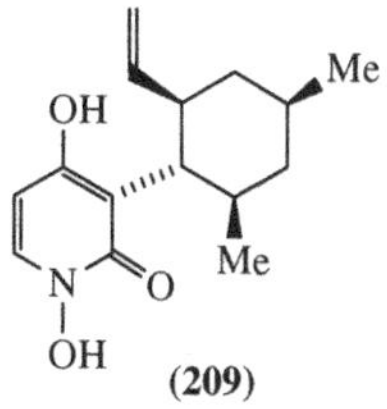

(209)
Pyridoxatin

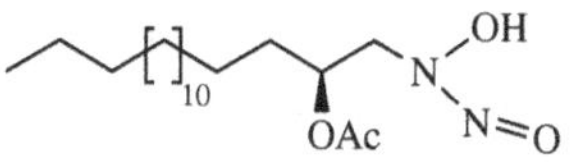

(210)
Poecillanosine

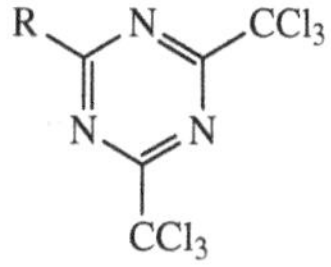

(211)

4. The formation of polymers by the photolysis of alkenes in the presence of an initiator has found many industrial applications. These include the formation of thin polymer films on printing plates and photoresists. Explain how triazines of type **(211)** can act as initiators for photopolymerizations.

5. Account for the following.
When a mixture of $CH_3N{=}NCH_3$ and $CD_3N{=}NCD_3$ is heated in the gas phase, CH_3CH_3, CH_3CD_3 and CD_3CD_3 are formed in a $1:2:1$ ratio. However, when the same reactants are heated in a solution of hydrocarbons, such as mineral (or paraffin) oil, only CH_3CH_3 and CD_3CD_3 are formed in equal amounts.

6. For the following reactions:

$$\text{EtOH} + CH_3CHO \xleftarrow{k_1} 2EtO^\bullet \xrightarrow{k_2} EtOOEt \qquad\qquad \begin{matrix} k_1/k_2 \\ 12.0 \end{matrix}$$

$$CH_3CH_3 + CH_2{=}CH_2 \xleftarrow{k_1} 2Et^\bullet \xrightarrow{k_2} CH_3CH_2CH_2CH_3 \qquad\qquad 0.15$$

(i) Give mechanisms to explain the formation of all the products derived from $EtO^\bullet$ and $Et^\bullet$.
(ii) Provide an explanation to account for the different k_1/k_2 ratios (where k is a rate constant).

7a. Continuous ultraviolet (UV) photolysis of a mixture of hydrogen peroxide and dimethyl ether gives an electron spin resonance (ESR) spectrum with three major peaks in the ratio $1:2:1$. (Each of these peaks is further split into a narrow $1:3:3:1$ quartet.) When dimethyl ether is replaced by diethyl ether, the ESR spectrum changes and eight major peaks are observed in the relative intensity $1 : 1 : 3 : 3 : 3 : 3 : 1 : 1$. (Each of these peaks is further split into a narrow $1 : 2 : 1$ triplet.)

(i) Give the structure of all the radicals which can be produced from the two reactions.

(ii) Highlight which of the radicals give rise to the ESR spectra and account for the appearance of the spectra.

7b. When the two reactions highlighted in part **7a** are carried out in the presence of the spin trap 2-methyl-2-nitrosopropane (MNP), the signals change. The dimethyl ether reaction produces an ESR spectrum with nine peaks, whereas the diethyl ether reaction produces a spectrum with six dominant peaks.

(i) Give the structure of the radicals which are now observed.

(ii) Account for the appearance of both ESR spectra.

8. Irradiation of a mixture of chlorine gas and (excess) pentanoyl chloride with UV light (at 35°C) leads to substitution of the hydrogen atoms for chlorine in the relative amounts (expressed as a percentage) given below.

$$CH_3-CH_2-CH_2-CH_2-COCl$$
14%   54%   29%   3%

Explain why the relative amounts of substitution at the four carbon atoms are different.

9. The relative rates of reaction of the *tert*-butyl radical with tetrahydrofuran, tris(trimethylsilyl)silane [$(Me_3Si)_3SiH$] and thiophenol (PhSH) at 25°C are $1 : 130 : 7500$, respectively.

(i) Draw the structures of the radicals which are formed from each of the three reactions.

(ii) Provide an explanation for the different reaction rates.

10. Account for the following.

(i) The alkoxyl radical $^tBu(Et)C(O^{\bullet})Me$ fragments to form almost exclusively one radical, even though fragmentation can occur by three different pathways.

(ii) The perester $PhCH=CHCH_2C(=O)-O-O-CMe_3$ decomposes several thousand times faster than the related perester $MeC(=O)-O-OCMe_3$ at the same temperature.

(iii) The *tert*-butyl radical reacts with acrylonitrile ($CH_2=CHCN$) around 3000 times faster than with 2-methylpropene (at the same temperature).

(iv) Addition of the phenyl radical (Ph$^\bullet$) to the benzene ring of benzonitrile (C_6H_5CN) or anisole ($C_6H_5OCH_3$) is faster than addition to benzene itself.

(v) Heating cyclohexene with molecular oxygen (at 55°C) leads to the regioselective formation of a hydroperoxide.

(vi) Reaction of 1,4-cyclohexadiene with primary alkyl radicals (R$^\bullet$) produces benzene and alkanes (RH) rather than products derived from addition to the double bond(s).

(vii) Photolysis of $CH_3C(=O)CH_2C(CH_3)_2CH(CH_3)_2$ produces 2,3-dimethyl-2-butene and propanone.

(viii) Photolysis of the hypochlorite, $CH_3CH_2CH_2CH_2CH_2CH_2OCl$, produces two different chloroalcohols (in a ratio of 10 : 1).

11. The rate constant for the (neophyl) rearrangement of radical (**212**) to radical (**213**) has a value of 1.1×10^3 s^{-1} at 25°C. This rearrangement has been used to estimate the rate constants of competing processes and is referred to as a 'clock reaction'. (For example, the rate constant for the reaction of (**212**) with Bu$_3$SnH can be determined from the ratio of products derived from the reaction of Bu$_3$SnH with (**212**) and (**213**), respectively, and the concentration of Bu$_3$SnH.) Provide a mechanism for the conversion of (**212**) to (**213**) and explain why the rearrangement takes place.

(212)   (213)

12. The natural product littorine (**214**) is known to rearrange to hyoscyamine (**215**) in the roots of certain plants. Mechanistic investigations have shown that, when (**214**) contains two ^{13}C atoms (represented by *), both of the ^{13}C labels are retained in (**215**). Suggest a mechanism for this rearrangement which is known to proceed via radical intermediates.

(214)   (215)

Part 2 (based on Chapters 5–11)

13. Three isomeric products, **(217)**–**(219)**, are obtained from the reaction of **(216)** as shown below. The product distribution is given for two different substituents, R = H or CO$_2$Me.

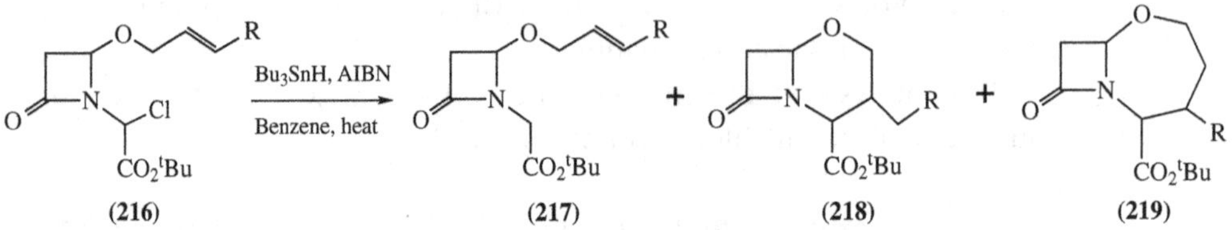

	(217) (%)	(218) (%)	(219) (%)
R = H	22	0	47
R = CO$_2$Me	16	68	4

(i) Provide reaction mechanisms to account for the formation of **(217)**–**(219)** from **(216)**.

(ii) Explain why changing the substituent R affects the ratio of **(218)** to **(219)**.

(iii) Which of **(218)** or **(219)** would you expect to be the major product when R = Ph? Explain your reasoning.

(iv) The reactions above involved the addition of Bu$_3$SnH and azobisisobutyronitrile (AIBN) to **(216)** in one portion. Would you expect a different product distribution if the reagents were added gradually to **(216)** using a syringe pump? Explain your reasoning.

14. Acid-catalysed hydrolysis of butyrolactone **(220)** has been used to prepare the food preservative sorbic acid, as shown below.

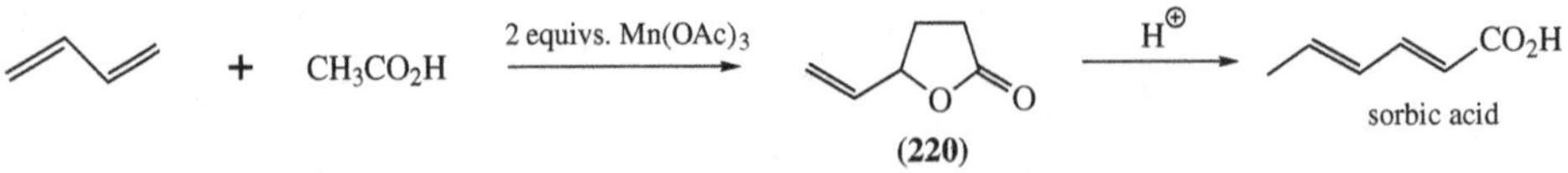

(i) **(220)** can be formed from reaction of butadiene and acetic (ethanoic) acid with 2 equivalents of manganese(III) acetate. What is the mechanism of this reaction?

(ii) For the synthesis of **(220)** on a large scale, the use of stoichiometric manganese(III) is not commercially viable. How could you modify the reaction conditions so that only catalytic quantities of manganese(III) are used?

15. Provide a mechanism to explain the stereoselective alkylation reaction shown below.

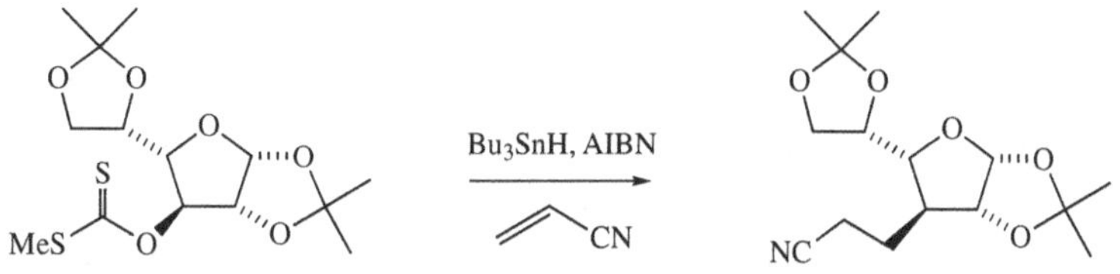

16. Provide a mechanism to explain the observed diastereoselectivity in the reaction below. What additional experiments could be carried out to provide further evidence for your proposed mechanism?

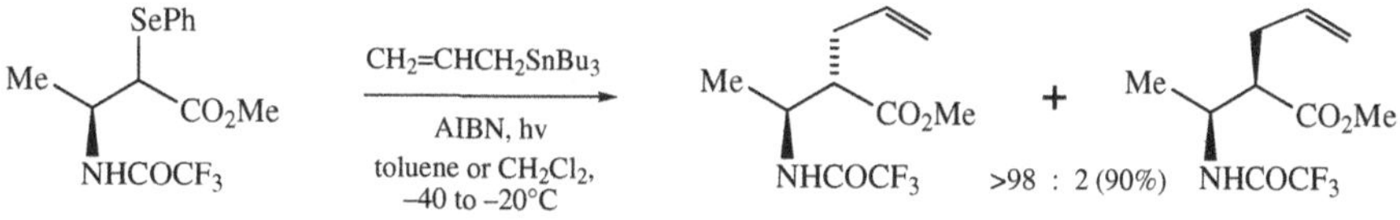

17. Provide a reaction mechanism to explain the following transformation. (Hint: the α-hydroxy group in the ketone is believed to play an important role in directing the stereoselectivity.)

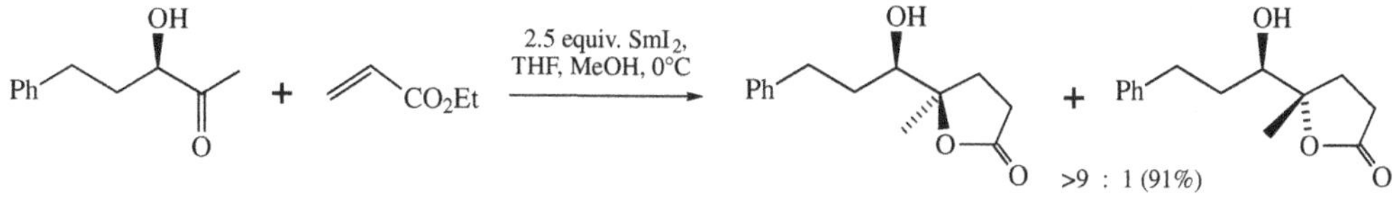

18. Provide reaction mechanisms to account for the following transformations.

BOC = *t*-butoxycarbonyl; PMB = *para*-methoxybenzyl

19. The biosynthesis of Amaryllidaceae alkaloids, such as **(222)** and **(223)**, is believed to involve the oxidative phenolic coupling of secondary amine **(221)**. Give all the possible products which can be formed from the oxidative phenolic coupling of **(221)**. Which of these can be converted to **(222)** and **(223)**?

(221)	**(222)**	**(223)**
	lycorine	haemanthamine

20. Radicals centred on atoms other than carbon can also undergo cyclization reactions. Propose a mechanism to explain the formation of tetracycle **(225)** from sulfenamide **(224)** on heating with Bu$_3$SnH/AIBN. [This involves the intermediacy of a nitrogen-centred (aminyl) radical.]

(224)	**(225)**	**(226)**	**(227)**	**(228)**

21. Tetrathiofulvalene (TTF) **(226)** is a well-known, one-electron reductant. It is used in materials chemistry to form stable radical ion pairs (on reaction with, for example, tetracyanoquinodimethane) which are electrical conductors. Propose a mechanism for the reaction of TTF with diazonium salt **(227)** to give the sulfonium salt **(228)**.

22. The weak P—H bonds (approximately 310 kJ mol^{-1}) in hypophosphorous acid **(229)** have led to the use of this compound as a radical-generating reagent in synthesis. Propose a mechanism for the reaction of diiodide **(230)** to give bi-cycle **(231)**, using **(229)** and AIBN in boiling aqueous sodium hydrogen carbonate. (Hint: **(229)** reacts in a similar manner to tributyltin hydride.)

(229)   (230)   (231)   (232)   (233)

23. Irradiation of 4-methylquinoline (lepidine) **(232)** in acidic methanol containing di-*tert*-butylperoxide gives rise to alkylation at the C-2 position. Give the product and outline a mechanism for its formation.

24. 2-Chloroquinoline **(233)** reacts with the anion derived from propanone (acetone) in the presence of Na(Hg)/NH$_3$ to give a product derived from C-2 substitution. Give the product and outline a mechanism for its formation.

25. Reaction of chlorobenzene with sodium trifluoroacetate at the anode of an electrochemical cell produces phenyl trifluoroacetate. Give a mechanism for this (oxidatively initiated) nucleophilic substitution reaction which involves an intermediate radical cation.

Outline Answers

Part 1 (based on Chapters 1–4)

1. Some of the decompositions could be concerted, rather than stepwise (as shown).

2. (i) $^tBuO^{\bullet}$ and $Me^{\bullet}$ produced on fragmentation ($+ \, ^tBuO^- + Fe^{III}$); reduction.

(ii) $PhCH_2C^{\bullet}(CO_2Et)_2$ ($+ \, Mn^{II}$); oxidation.

(iii) $Ph^{\bullet}$ ($+ \, Cu^{II} + N_2$); reduction.

(iv) $EtCO_2^{\bullet}$ and $Et^{\bullet}$ (on loss of CO_2); oxidation.

(v) $PhO^{\bullet}$ ($+ \, H^+ + Ce^{III}$); oxidation.

(vi) $CH_3OC_6H_4CH_2^{\bullet}$ (from reaction with Ag^{II}; $+ \, H^+$); oxidation.

3. Carazostatin: the O—H bond on the benzene ring is weak, and therefore reaction with radicals can lead to hydrogen atom abstraction to give a resonance-stabilized phenoxyl radical. [Choshi, T., Sada, T., Fujimoto, H., *et al.* (1996) *Tetrahedron Letters*, **37**, 2593–2596.]

Pyridoxatin: this has two weak O—H bonds, one on the benzene ring and the second on nitrogen. Abstraction of the NO—H hydrogen atom would lead to a resonance-stabilized nitroxide radical. [Snider, B.B. & Lu, Q. (1994) *Journal of Organic Chemistry*, **59**, 8065–8070.]

Poecillanosine: this has a weak NO—H bond and also a nitroso group to which radicals can add, leading to nitroxides [cf. nitroso spin traps in electron spin resonance (ESR)]. [Natori, T., Kataoka, Y., Kato, S., Kawai, H. & Fusetani, N. (1997) *Tetrahedron Letters*, **38**, 8349–8350.]

4. The weakest bond of the triazine is the C—Cl bond. Photolysis of this bond forms a resonance-stabilized radical which can add to alkenes leading to polymerization. As two CCl_3 groups are present, the polymer could be initiated from either or both of these sites.

5. In the gas phase, the $^{\bullet}CH_3$ and $^{\bullet}CD_3$ radicals (generated on decomposition of the azoalkanes) will combine with a diffusion-controlled rate to form all three possible coupling products in a statistical ratio. However, in viscous (hydrocarbon) solvents, a solvent cage surrounds the radicals (and nitrogen), preventing their escape. This leads to exclusive homocoupling of $^{\bullet}CH_3$ and $^{\bullet}CD_3$ within the cage.

6. (i)

(ii) The alkoxyl radical ($EtO^{\bullet}$) favours disproportionation (over combination) because this leads to the formation of a strong C=O bond, while combination forms a much weaker O—O (peroxide) bond. In contrast, disproportionation of the alkyl radical ($Et^{\bullet}$) leads to a C=C bond which is weaker than C=O, and therefore combination to form a C—C bond, which is stronger than O—O, is predominantly observed.

7a. (i) $HO^{\bullet}$, $CH_3OCH_2^{\bullet}$, $CH_3CH_2OCH_2CH_2^{\bullet}$, $CH_3CH_2OCH^{\bullet}CH_3$ ($CH_3^{\bullet}$ and $CH_3CH_2^{\bullet}$ could also be formed on radical fragmentation).
(ii) $CH_3OCH_2^{\bullet}$ gives rise to the triplet due to signal splitting by two α-hydrogens (the quartet is due to long-range splitting by the CH_3 group). $CH_3CH_2OCH^{\bullet}CH_3$ gives rise to the eight-line signal due to splitting by one α-hydrogen (to give a doublet) and further splitting by three β-hydrogens. (Further splitting to give triplets is due to long-range splitting by the CH_2 group.)

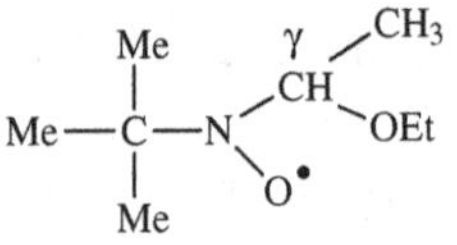

7b. (i) and (ii)

9 peaks: peak split into 3 lines by nitrogen and each of these is further split by two γ-CH_2 hydrogens

6 dominant peaks: peak split into 3 lines by nitrogen and each of these is split by one γ-CH hydrogen

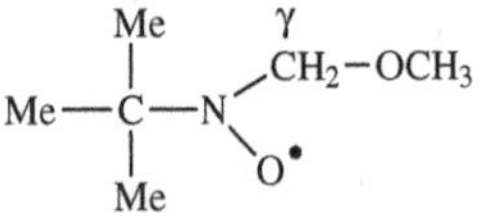

8. The chlorine radical is electrophilic and therefore will preferentially attack sites of higher electron density. Therefore, the further the methylene group is from the (electron-withdrawing) acid chloride group, the greater the chlorination. The low percentage of chlorination at the methyl group is due to the primary C—H bonds being stronger than the secondary C—H bonds (hence primary radicals are less stable than secondary radicals).

9. (i)

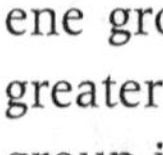

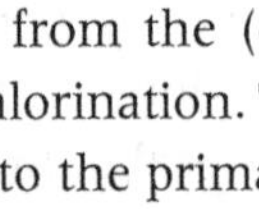

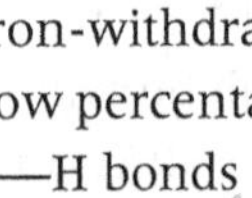

(ii) The different reaction rates can be explained by polar effects. As the *tert*-butyl radical is nucleophilic, it will react most rapidly with sites of lowest electron density. Hence, hydrogen atom abstraction of the electron-poor hydrogen of the thiol ($PhS^{\delta-}$—$H^{\delta+}$), to give an electrophilic thiyl radical ($PhS^{\bullet}$), is very fast. In contrast, the hydrogen atoms which are abstracted in tris(trimethylsilyl)silane ($Si^{\delta+}$—$H^{\delta-}$) and tetrahydrofuran are more electron rich and these reactions produce nucleophilic radicals.

10. (i) Fragmentation to give the $^{t}Bu^{\bullet}$ radical is favoured because this (tertiary radical) is more stable than $Et^{\bullet}$ or $Me^{\bullet}$.

(ii) Homolysis of the weak O—O bond produces two different carboxyl radicals (and $Me_3CO^{\bullet}$) which can undergo decarboxylation. The difference in the rate of decomposition reflects the increased stability of the $PhCH{=}CHCH_2^{\bullet}$ radical (which is stabilized by resonance) over the $CH_3^{\bullet}$ radical.

(iii) As the *tert*-butyl radical is nucleophilic, it has a high-energy singly occupied molecular orbital (SOMO) and therefore reacts fastest with electron-poor double bonds, such as that in acrylonitrile, which have low-energy lowest unoccupied molecular orbitals (LUMOs). (The greater the SOMO–LUMO interaction, the greater the reactivity.)

(iv) The introduction of electron-rich (OMe) or electron-poor (CN) substituents onto the benzene ring increases the rate of radical addition because the cyclohexadienyl radicals can be (resonance) stabilized by the OMe or CN substituents.

(v) Autoxidation occurs selectively at the allylic position (adjacent to the C=C bond) because the C—H bonds at this position are weaker and more readily attacked. Hydrogen atom abstraction (presumably by peroxyl radicals, $ROO^{\bullet}$) produces an allylic radical which is stabilized by resonance, and this reacts with molecular oxygen to form a hydroperoxide.

(vi) The doubly allylic C—H bonds of 1,4-cyclohexadiene are very weak (approximately 305 kJ mol^{-1}) because the resultant radical is stabilized by resonance. Hence, hydrogen atom abstraction is preferred. Reaction of the cyclohexadienyl radical with a further radical leads to the loss of a second hydrogen atom to give benzene, which is stabilized by aromaticity.

(vii) This is an example of a Norrish type II reaction which proceeds via a selective 1,5-hydrogen atom abstraction.

(viii) Photolysis produces an alkoxyl radical which gives $CH_3CH_2CH(Cl)CH_2CH_2CH_2OH$ and $CH_3CH(Cl)CH_2CH_2CH_2CH_2OH$ on 1,5- and 1,6-hydrogen atom transfer, respectively, followed by chlorine atom abstraction. (The predominant product arises from 1,5-hydrogen atom transfer.)

11.

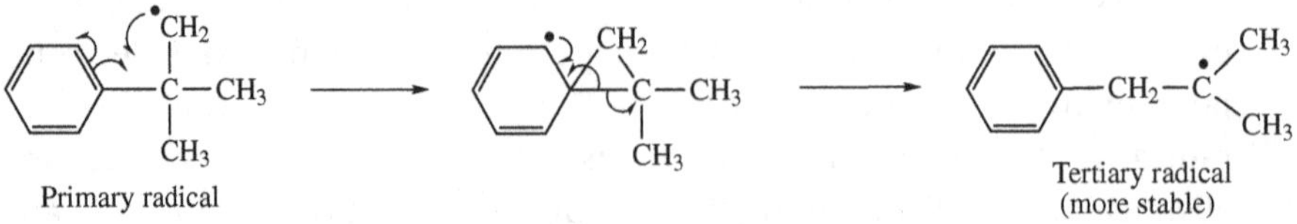

Primary radical   Tertiary radical (more stable)

12.

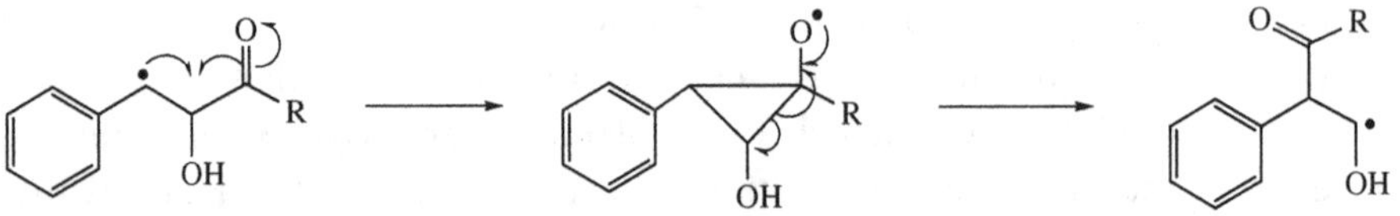

(R.B. Herbert, *Nat. Prod. Rep.*, 1997, **14**, 359–372)

Part 2 (based on Chapters 5–11)

13. (i)

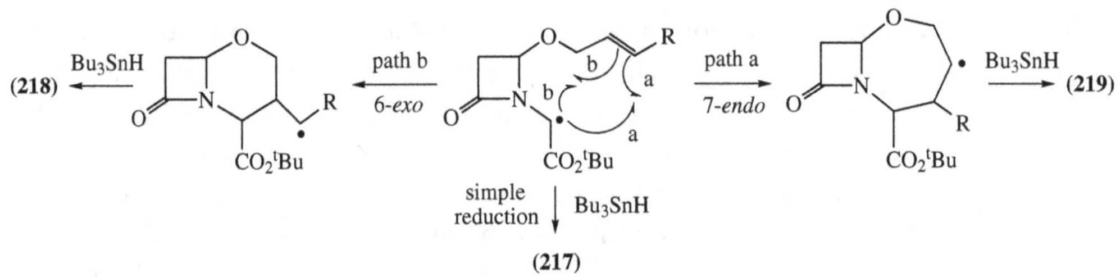

(ii) These cyclizations are likely to be reversible and under thermodynamic control because of the stability of the initial (captodative) radical generated on chlorine atom abstraction from the precursor. When R = H, radical addition to the least hindered carbon atom of the alkene, to give the more substituted radical, is favoured. However, when R = CO_2Me, steric and electronic effects favour the formation of a six-membered ring. Thus, the size of the ester group disfavours attack to give the seven-membered ring, while the radical derived from 6-*exo* cyclization can be stabilized by the ester group (by resonance).

(iii) **(218)**, is expected because this would involve an intermediate (resonance-stabilized) benzylic radical.

(iv) More cyclization, to give **(218)** and **(219)**, is expected over simple reduction (to give **217**). Slow addition of Bu_3SnH would allow the initial captodative radical more time for cyclization. [Bachi, M.D., Frolow, F. & Hoornaert, C. (1983) *Journal of Organic Chemistry*, **48**, 1841–1849.]

14. (i)

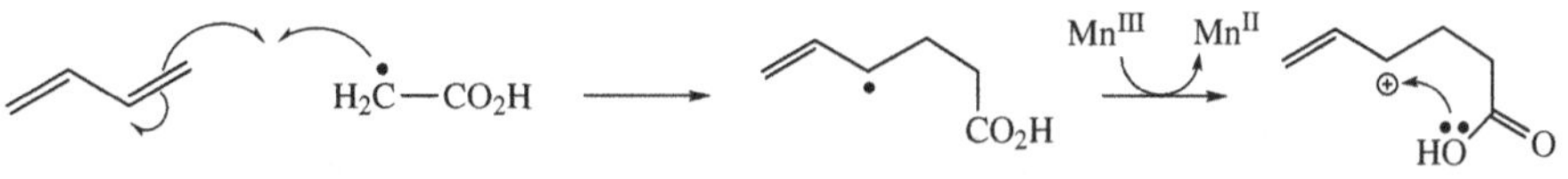

(ii) Manganese(II) needs to be oxidized to manganese(III). This has been carried out on a large scale by electrochemical oxidation at the anode. [Coleman, J.P., Hallcher, R.C., McMackins, D.E., Rogers, T.E. & Wagenknecht, J.H. (1991) *Tetrahedron*, **47**, 809–829.]

15.

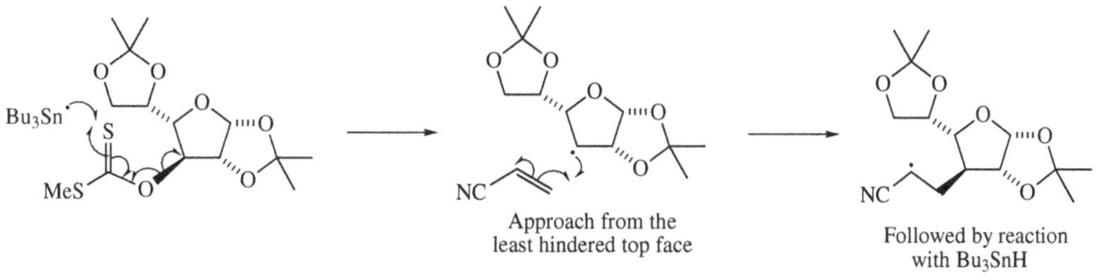

(B. Giese, J.A. Gonzales-Gomez and T. Witzel, *Angew. Chem.*, 1984, 96, 51)

16.

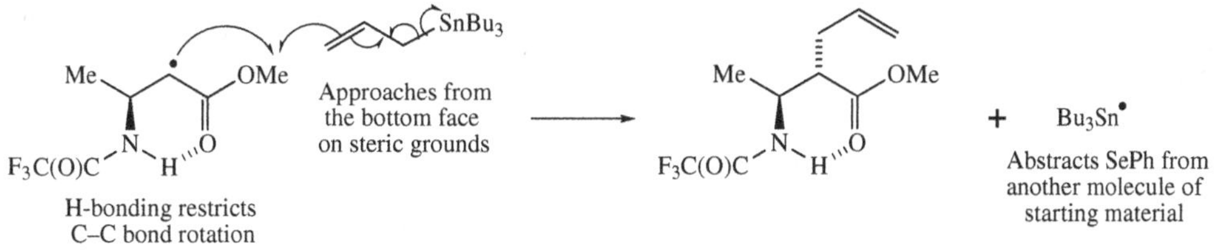

Carrying out the reaction in a solvent which can form intermolecular hydrogen bonds with the reactant, such as dimethylsulfoxide [$CH_3S(O)CH_3$], will break the intramolecular hydrogen bond leading to free rotation and erosion or reversal of the diastereoselectivity. Alternatively, replacing the hydrogen on nitrogen with, for example, an alkyl group will prevent any intramolecular hydrogen bonding. [Hanessian, S., Yang, H. & Schaum, R. (1996) *Journal of the American Chemical Society*, **118**, 2507–2508.]

17.

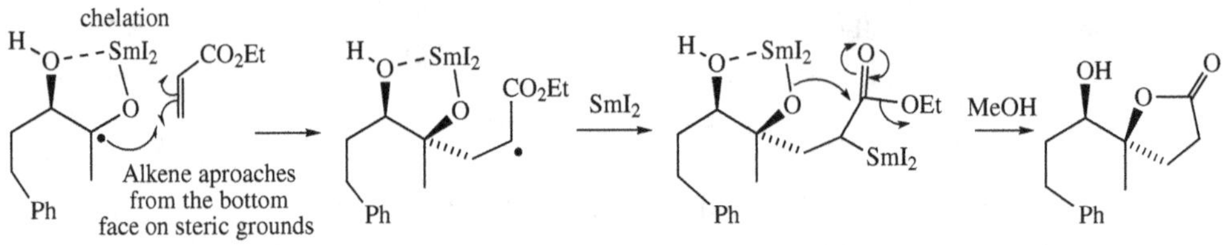

(M. Kawatsura, F. Matsuda and H. Shirahama, *J. Org. Chem.*, 1994, **59**, 6900–6901).

18.

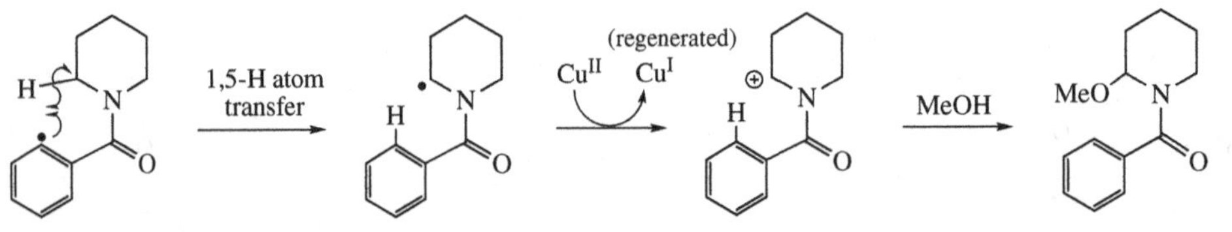

(G. Han, M.C. Mcintosh amd S.M. Weinreb, *Tetrahedron Lett.*, 1994, **35**, 5813–5816).

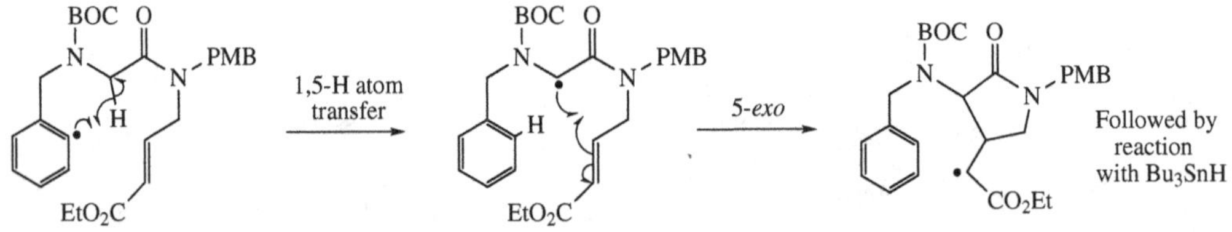

(J. Rancourt, V. Gorys and E. Jolicoeur, *Tetrahedron Lett.*, 1998, **39**, 5339–5342).

19.

MeO
3
2
H
N
5
6
O
1
4
O

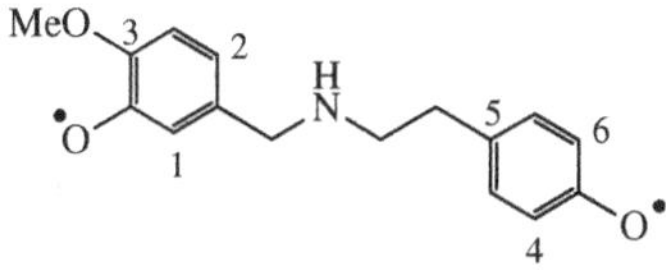

Coupling to form C−C bonds could involve radical combination at positions 1–3 with either of 4–6.
Coupling at positions 2 and 4 (or 6) is thought to be involved in the biosynthesis of **222**.
Coupling at positions 2 and 5 is thought to be involved in the biosynthesis of **223**.

20.

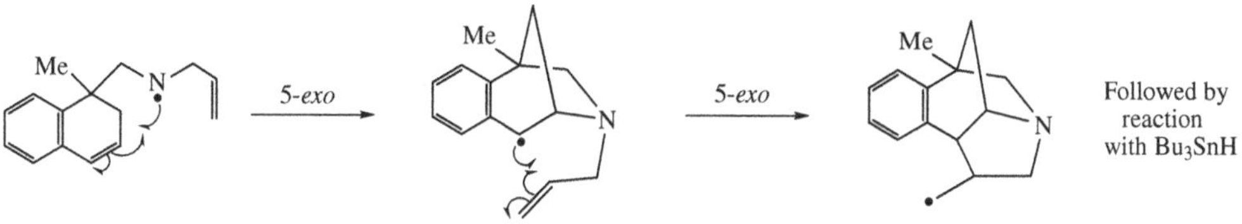

Me
N
5-*exo*
Me
N
5-*exo*
Me
N
Followed by reaction with Bu_3SnH

(W. R. Bowman, D. N. Clark and R.J. Marmon, *Tetrahedron*, 1994, **50**, 1295-1310).

21.

N_2 +
S
S
S
S
BF_4
+
O
5-*exo*
O
+
S
S
S
S

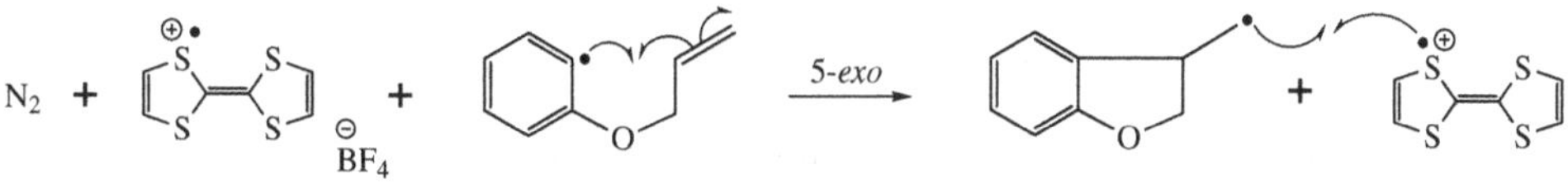

(R. J. Fletcher, C. Lampard, J.A. Murphy and N. Lewis, *J.Chem. Soc., Perkin Trans. 1*, 1995, 623–633).

22.

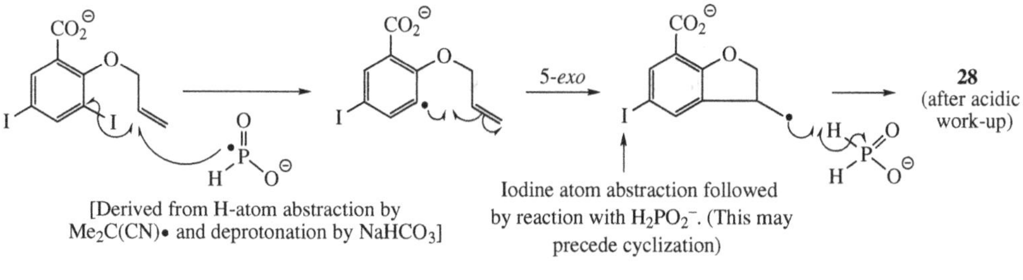

CO_2
O
I
I
O
P
H
O
[Derived from H-atom abstraction by $Me_2C(CN)$• and deprotonation by $NaHCO_3$]
CO_2
O
I
5-*exo*
CO_2
O
I
Iodine atom abstraction followed by reaction with $H_2PO_2^-$. (This may precede cyclization)
H
O
P
H
O
28
(after acidic work-up)

(S. R. Graham, J.A. Murphy and D. Coates, *Tetrahedron Lett*, 1999, **40**, 2415–2416).

23.

(Derived from reaction
of MeOH with ^{t}BuO•)

(W. Buratti, G.P. Garcini, F. Minisci, F. Bertini, R. Galli and M. Perchinunno, *Tetrahedron*, 1971, **27**, 3655–3668).

24. This is a unimolecular radical nucleophilic substitution ($S_{RN}1$) reaction.

Formed on electron
transfer from Na

electron transfer
to 2-chloroquinoline

(E. Austin, C.G. Ferrayoli, R.A. Alonso and R.A. Rossi, *Tetrahedron*, 1993, **49**, 4495–4502).

25. This is an example of an $S_{ON}2$ reaction (substitution, oxidative, nucleo-philic, bimolecular).

(L. Eberson, L. Jönsson and L.G. Wistrand, *Tetrahedron*, 1982, **38**, 1087–1093).

Further Reading

Alfesi, Z.B. (1999) *General Aspects of the Chemistry of Radicals*. Wiley, New York.

Bauld, N.L. (1997) *Radicals, Ion Radicals and Triplets*. Wiley-VCH, New York.

Curran, D.P., Porter, N.A. & Giese, B. (1996) *Stereochemistry of Radical Reactions. Concepts, Guidelines and Synthetic Applications*. VCH, Weinheim.

Fischer, H. (ed.) (1983) *Radical Reaction Rates in Liquids*, Landolt-Börnstein, New Series, Vol. 13. Springer Verlag, Berlin.

Fossey, J., Lefort, D. & Sorba, J. (1995) *Free Radicals in Organic Synthesis*. Wiley, New York.

Giese, B. (1986) *Radicals in Organic Synthesis: Formation of Carbon–Carbon Bonds*. Pergamon, Oxford.

Kochi, J.K. (1973) *Free Radicals*, Vols 1 and 2. Wiley, New York.

Moody, C.J. & Whitham, G.H. (1992) *Reactive Intermediates*. Oxford University Press, Oxford.

Motherwell, W.B. & Crich, D. (1992) *Free Radical Chain Reactions in Organic Synthesis*. Academic Press, London.

Nonhebel, D.C., Tedder, J.M. & Walton, J.C. (1979) *Radicals*. Cambridge University Press, Cambridge.

Perkins, M.J. (1994) *Radical Chemistry*. Ellis Horwood, New York.

Perkins, M.J. (2000) *Radical Chemistry – the Fundamentals*. Oxford University Press, New York.

Wentrup, C. (1984) *Reactive Molecules*. Wiley, New York.

Index

Page references to figures appear in *italic* type and those for tables appear in **bold** type.